T0181010

Task Scheduling for Multi-core and Parallel Architectures

Quan Chen · Minyi Guo

Task Scheduling for Multi-core and Parallel Architectures

Challenges, Solutions and Perspectives

Quan Chen
Shanghai Jiao Tong University
Shanghai
China

Minyi Guo
Shanghai Jiao Tong University
Shanghai
China

ISBN 978-981-13-4835-8 ISBN 978-981-10-6238-4 (eBook)
https://doi.org/10.1007/978-981-10-6238-4

© Springer Nature Singapore Pte Ltd. 2017
Softcover re-print of the Hardcover 1st edition 2017
This work is subject to copyright. All rights are reserved by the Publisher, whether the whole or part of the material is concerned, specifically the rights of translation, reprinting, reuse of illustrations, recitation, broadcasting, reproduction on microfilms or in any other physical way, and transmission or information storage and retrieval, electronic adaptation, computer software, or by similar or dissimilar methodology now known or hereafter developed.
The use of general descriptive names, registered names, trademarks, service marks, etc. in this publication does not imply, even in the absence of a specific statement, that such names are exempt from the relevant protective laws and regulations and therefore free for general use.
The publisher, the authors and the editors are safe to assume that the advice and information in this book are believed to be true and accurate at the date of publication. Neither the publisher nor the authors or the editors give a warranty, express or implied, with respect to the material contained herein or for any errors or omissions that may have been made. The publisher remains neutral with regard to jurisdictional claims in published maps and institutional affiliations.

Printed on acid-free paper

This Springer imprint is published by Springer Nature
The registered company is Springer Nature Singapore Pte Ltd.
The registered company address is: 152 Beach Road, #21-01/04 Gateway East, Singapore 189721, Singapore

Preface

Computers play an indispensable role in almost all scientific fields and disciplines such as biomedicine, physical simulations, computational chemistry, and astronautics. In order to fulfill the urgent requirement for high computational capacity (for example, huge amount of computational capacity is required in big data analysis), technologies of parallel computing and resource management increasingly grow. Nowadays, multi-core processors have become mainstream in both research and real-world settings, from warehouse-scale datacenter, personal desktops, laptops, to smartphones, since they demonstrate the superior performance per watt and the larger computational capacity compared to the traditional single-core processors. For these widely used parallel systems, a key problem is how to schedule the tasks to efficiently utilize the hardware, improve the performance, and guarantee the Quality-of-Service. Because different types of architectures have totally different features, there is not universal scheduling technique that can work the best across all these architectures and different applications. Scheduling techniques often need to be altered to match the various architectures (e.g., multi-core, datacenter, and distributed system) accordingly. For example, for a big data processing system that runs on large-scale distributed system, task scheduling should focus on load balancing and data locality; for a datacenter, the scheduling should aim to guarantee the quality-of-service of the customer-facing applications (e.g., web search).

However, a book that elaborates task scheduling techniques for emerging complex parallel architectures (e.g., multi-socket architecture, heterogeneous multi-core architecture, and cloud/big data processing platform) is missing. Previous published works mainly discuss traditional task scheduling techniques for generalized parallel systems mathematically. But these techniques suffer from low performance on the emerging complex parallel architectures. To this end, this book discusses the state-of-the-art task scheduling techniques that are optimized against different architectures, and these techniques can be applied in real parallel systems directly.

In this book, we will mainly introduce these task scheduling techniques in the scenarios of emerging parallel architectures, including the multi-core architecture, cloud platform, and accelerator. The book will examine the current challenges in this topic, and present detailed solutions including algorithms, methods, and perspectives. It is well noticed that parallel architectures are becoming more and more complex. Instances are multi-socket multi-core architecture, asymmetric multi-core architecture, various parallel accelerators, distributed parallel architecture, etc. For different types of parallel architectures, in order to utilize the hardware efficiently and maximize the performance, different techniques have been introduced and integrated into the traditional task scheduling policy. To this end, we elaborate these techniques and demonstrate that they can be implemented efficiently for emerging parallel architectures. This book consists of three main parts: *background*, *task scheduling techniques*, and *perspectives*.

Part I: Background

In this part, we mainly introduce the background of this book, including the emerging widely used parallel architectures and classic task scheduling techniques. This part includes two chapters.

In the first chapter, we do a survey on the emerging parallel architectures including multi-core architectures, NUMA-enabled multi-core architectures, asymmetric multi-core architectures, accelerators, cloud platform, and so on. Besides these parallel architectures, vendors are now producing other architectures. For instance, Google releases Tensor Processing Unit (TPU), which is a parallel architecture as well recently.

In the second chapter, we introduce the classic task scheduling policies and parallel programming environments. Work-sharing and work-stealing are the two most classic task scheduling policies. In addition, we introduce many parallel programming environments, such as Apache Hadoop, Spark, MIT Cilk, TBB, X10, and so on. These task scheduling policies can be applied in various parallel architectures but may suffer from low performance. In the next part, we introduce the techniques that optimize the parallel applications on various parallel architectures.

Part II: Task Scheduling for Various Parallel Architectures

In this part, we introduce techniques that can be used to improve the performance of applications on the emerging widely used parallel architectures.

In the third chapter, targeting the multi-socket architecture, we will introduce cache-aware task scheduling policy, which can improve shared cache utilization in different sockets.

In the fourth chapter, on the NUMA (Non-Uniform Memory Access)-enabled architecture, the NUMA-aware task scheduling policy will be introduced, which can reduce remote memory accesses and improve the performance of applications in consequence.

In the fifth chapter, we introduce workload-aware task scheduling policy, which is proposed for asymmetric multi-core architecture. On this kind of architecture, where different cores operate at different speeds, partitioning by the workload can truly improve the performance.

In the sixth chapter, we introduce asymptotic technique to allocate workload between CPU and CPU for CPU+GPU heterogeneous parallel architecture. On this kind of architecture, different applications have different speedup ratios on the GPU compared with CPU, because the applications have various characteristics. It is not trivial to find an optimal workload partition between CPU and GPU.

Nowadays, big data analysis requires tremendous amount of computers to process the data in parallel. In the seventh chapter, we will introduce several featured dynamic task scheduling policies that can significantly improve the performance of big data processing on heterogeneous cloud platforms.

The eighth chapter contains the quality-of-service aware task scheduling policy for accelerators. Using this policy, it can improve the accelerator utilization. Moreover, it also guarantees the quality-of-service of latency-critical applications.

Part III: Summary and Perspectives

In this part, we summarize all the previously introduced task scheduling solutions for parallel architectures, provide our perspectives, and discuss the possibilities of designing new dynamic task scheduling policies for more other future parallel architectures. Especially, we give several guidelines of designing new efficient and effective task scheduling techniques for those newly released parallel architectures.

After reading this book, we expect the readers will have an overview on the recent progress of task scheduling policies in parallel architectures. And we also hope the book can help the readers to quickly master the focused issues and opening problems if they tend to work in this field. In order to understand this book, the readers are suggested to have some basic knowledge on computer architecture, multi-core, and parallel processing.

Shanghai, China
September 2017

Quan Chen
Minyi Guo

Acknowledgements

This book was partially sponsored by the National Basic Research 973 Program of China under grant 2015CB352403, the National Natural Science Foundation of China (NSFC) (61602301). We are grateful for the editor of this book, Dr. XiaoLan Yao at Springer for her patience and support to make this book possible.

Shanghai, China
September 2017

Quan Chen
Minyi Guo

Contents

Acronyms

AATS	Asymmetric-Aware Task Scheduling
AMC	Asymmetric Multi-Core
ANN	Approximate Nearest Neighbor
BIOS	Basic Input/Output System
CAB	Cache-Aware Bi-tier Work-Stealing
CF	Cache-Friendly
c-group	Core Group
CMPI	Cache Misses Per Instruction
CPU	Central Processing Unit
D&C	Divide-and-Conquer
DAG	Directed Acyclic Graph
DDR	Dual Data Rate
DRAM	Dynamic Random Access Memory
DT	Duration Table
DVFS	Dynamic Voltage and Frequency Scaling
EDC	MCDRAM Controller
FIFO	First-In-First-Out
FPGA	Field-Programmable Gate Array
FTO	Full Tree Oriented
GFS	Google File System
GPGPU	General-Purpose Graph Processing Unit
GTO	General Tree Oriented
HATS	Heterogeneous-Aware Task Scheduler
HDFS	Hadoop Distributed File System
HTT	Hyper-Threading Technology
IC	Inter-Connect
IPA	Intelligent Personal Assistant
ISA	Instruction Set Architecture
KNL	Intel Xeon Phi (codenamed Knights Landing)
KNN	K-Nearest Neighbor

LATE	Longest Approximate Time to End
LAWS	Locality-Aware Work-Stealing
LR	Linear Regression
MCDRAM	Multi-Channel Dynamic Random Access Memory
MIMD	Multiple-Instruction-Multiple-Data
MPI	Message Passing Interface
MPS	Multi-Process Service
MSMC	Multi-Socket Multi-core
NUMA	Non-Uniform Memory Access
PFWS	Parent-First Work-Stealing
PMC	Performance Monitoring Counter
QoS	Quality-of-Service
RDD	Resilient Distributed Dataset, defined in Apache Spark
RPC	Remote Procedure Call
SAMR	Self-Adaptive Map-Reduce
SIMD	Single-Instruction-Multiple-Data
SM	Streaming Multiprocessor
SMC	Symmetric Multi-core
SOID	Size Of Involved Data
SVM	Support Vector Machines
TBB	Intel Thread Building Blocks
TCO	Total Cost of Ownership
TRICI	Task Relocation Incurred Cache Interference problem

Part I
Background

Chapter 1
Emerging Parallel Architectures

Abstract In this chapter, a survey on emerging parallel architectures including Multi-core architectures, accelerators, Datacenter, and distributed system, and the state-of-the-art dynamic task scheduling systems will be presented respectively. According to the category of the architecture, in the following chapters, we will introduce the detailed techniques of task scheduling that can be efficiently used on the different parallel architectures.

1.1 Parallel Architecture is Dominating the World

Computers play an indispensable role in almost all scientific fields and disciplines including biomedicine, physical simulations, computational chemistry, aeronautics, and astronautics. In order to fulfill the urgent requirement for high computational capacity, emerging computer technology demands increasingly on parallel computing, such as multi-core architecture, which integrates multiple cores in a central processing unit (CPU). Nowadays, multicore processors have become mainstream in both research and public settings, from supercomputers to personal laptops to smartphones, since they demonstrate superior performance per watt and larger computational capacity than single-core processors.

Traditionally, there are generally two main categories of parallel architectures: *shared memory parallel architecture* and *distributed memory parallel architecture*. In shared memory parallel architecture, the main memory is shared by all the processing elements (e.g., cores in CPU); On the other hand, in distributed memory parallel architecture, the main memory is distributed to different nodes and each processing element can only access part of the main memory. In the past decade, parallel architecture evolves quickly, and some new types of parallel accelerators, such as General Purpose Graph Processing Unit (GPGPU), Intel Xeon Phi, are used in actual computers. In the following several sections, we introduce these emerging popular parallel architectures respectively. In more detail, in Sect. 1.2, we introduce shared memory parallel architecture. In Sect. 1.3, we introduce distributed memory parallel architecture. In Sect. 1.4, we introduce the parallel accelerator architecture.

© Springer Nature Singapore Pte Ltd. 2017

Q. Chen and M. Guo, *Task Scheduling for Multi-core and Parallel Architectures*, https://doi.org/10.1007/978-981-10-6238-4_1

When assigning tasks to processing elements in parallel architectures, it is crucial to balance workloads among all the cores so that the parallel architectures can be utilized most effectively. To achieve this purpose, researchers have proposed a large amount of efficient task scheduling policies. We introduce popular task scheduling policies in Chap. 2.

1.2 Shared Memory Parallel Architecture

In shared memory parallel architecture, the main memory is shared by all the processing elements (cores), and every processing element can access data stored in the shared memory directly. Without loss of generality, in the following of this book, we use cores to represent processing elements.

1.2.1 Multi-core Architecture

Multi-core architecture, where multiple cores are integrated in the same CPU, is the most classic shared memory parallel architecture. Figure 1.1 presents an example of multi-core architecture.

As shown in the figure, each core has its own private cache, and all the cores share the last level cache. In addition, all the cores can access all the data stored in the main memory directly. Meanwhile, all the cores have the same latency to access the data from the shared memory. Compared with traditional single-core architecture, multi-core architecture provides much high computational ability and is much more energy efficient.

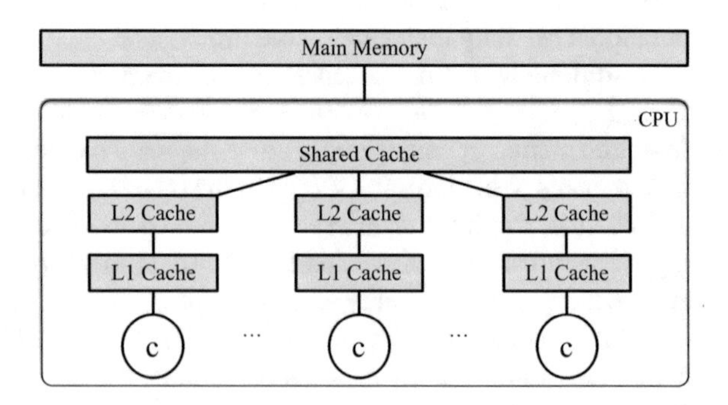

Fig. 1.1 Classic multi-core architecture

1.2.2 Multi-socket Multi-core Architecture

Although hardware manufacturers keep increasing cores in CPU chips, the number of cores cannot be increased unlimitedly due to physical limitations. To meet the urgent need for powerful computers, multiple CPU chips are integrated into a Multi-Socket Multi-Core (MSMC) architecture, in which each CPU chip has multiple cores with a shared last-level cache and is plugged into a socket. MSMC architecture has already widely adopted in emerging supercomputers and clusters.

Meanwhile, modern shared-memory MSMC computers and large-scale supercomputing systems often employ NUMA-based (*Non-Uniform Memory Access*) memory system, in which the whole main memory is divided into multiple memory nodes and each node is attached to the socket of a chip. The memory node attached to a socket is called its local memory node and those that are attached to other sockets are called remote memory nodes. The cores of a socket access its local memory node much faster than the remote memory nodes.

Figure 1.2 gives an example of MSMC computers that employ NUMA-based memory system. As shown in the figure, different sockets are connected through *inter-connect* (IC) modules. For example, in Intel-based machine, QPI [7] is used to connect different sockets, while in AMD-based machine, HyperTransport [3] is used to connect different sockets. While a core c in CPU S accesses (either read or write) data from the memory node attached to another socket S_r, the data is transferred from S_r's memory node to core c through IC modules that connect socket S and socket S_r.

Multi-socket multi-core architecture is developed on the basis of multi-core architecture. The main difference between MSMC architecture and traditional multi-core architecture is that, cores have the same latency to access data from memory in tra-

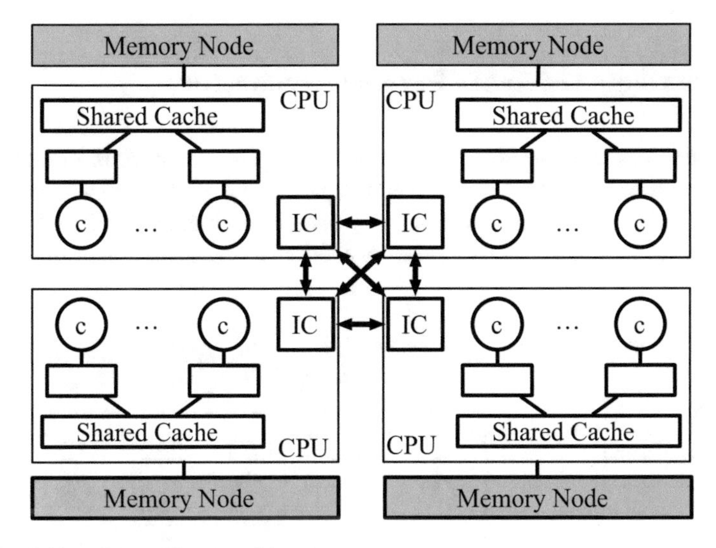

Fig. 1.2 Multi-socket multi-core architecture

ditional multi-core architecture, while have different latencies to access data from local memory node and remote memory nodes in MSMC architecture.

1.2.3 Asymmetric Multi-core Architecture

While chip manufacturers like AMD and Intel keep producing new CPU chips with more symmetric cores, researchers are investigating alternative multi-core organizations such as Asymmetric Multi-Core (AMC) architectures, where individual cores have different computational capabilities [1–4].

AMC is attractive because it has the potential to improve system performance, to reduce power consumption, and to mitigate Amdahl's law [1, 4]. Since an AMC architecture consists of a mix of fast cores and slow cores, it can better cater for applications with a heterogeneous mix of workloads [2, 3]. For example, fast, complex cores can be used to execute the serial code sections, while slow, simple cores can be used to crunch numbers in parallel, which is more power-efficient. For example, Nintendo WII and Nintendo DS use AMC processors. Also, many modern multi-core chips offer Dynamic Voltage and Frequency Scaling (DVFS) which can dynamically adjust the operating frequency of each core and thus is able to turn a symmetric multi-core chip into a performance-asymmetric multi-core chip.

Figure 1.3 presents an example of asymmetric multi-core architectures. In AMC architectures, such as the Intel Quick-IA [2] and ARM Big-Little [5] architectures, different types of cores have different computational capacities. On the contrary, in traditional multi-core architecture, all the cores have the same computational capacities.

1.3 Distributed Memory Parallel Architecture

Large scale clusters adopt distributed memory parallel architecture in most cases. While the memory is shared between all the processing elements in shared memory parallel architecture, in distributed memory parallel architecture, different computer nodes often have their own private memory. When a core needs data stored in other

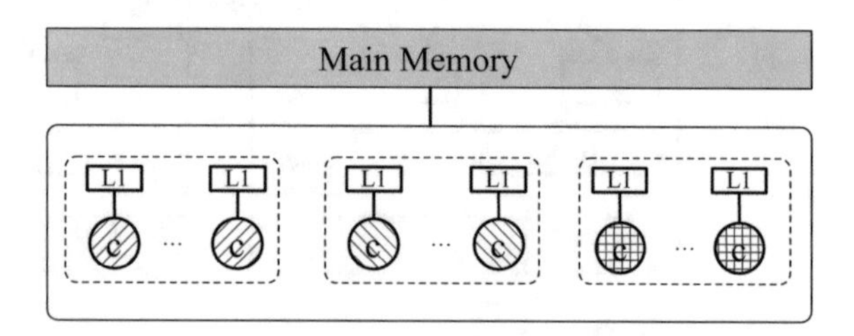

Fig. 1.3 Asymmetric multi-core architecture. In this figure, cores with different fill patterns have different computation capacities

nodes' memory, the data is transferred through networks. According to how the computer nodes are connected, there are *tight-coupled distributed memory architecture* and *loose-coupled distributed memory architecture*.

1.3.1 Tight-Coupled Distributed Memory Architecture

In tight-coupled distributed memory architecture, the computer nodes are connected by local high-speed inter-connect network using high speed Ethernet and InfiniBand. Traditional *clusters* and emerging *datacenters* adopt tight-coupled distributed memory architecture. These high performance distributed memory platforms are used to house Cloud computing or large scale web applications.

Figure 1.4 presents how clouds are built on top of datacenters. It is worthing noting that, the Compute resource and data resource can locate on the same computer node. In public Cloud, once the resources are leased/rented by users, the users are responsible to manage the resources themselves.

1.3.2 Loose-Coupled Distributed Memory Architecture

In contrast to tight-coupled distributed memory architecture, in loose-coupled distributed memory architecture, computer nodes are normally connected with conventional network interface, such as Internet. *Grid computing* [4] is one of the most

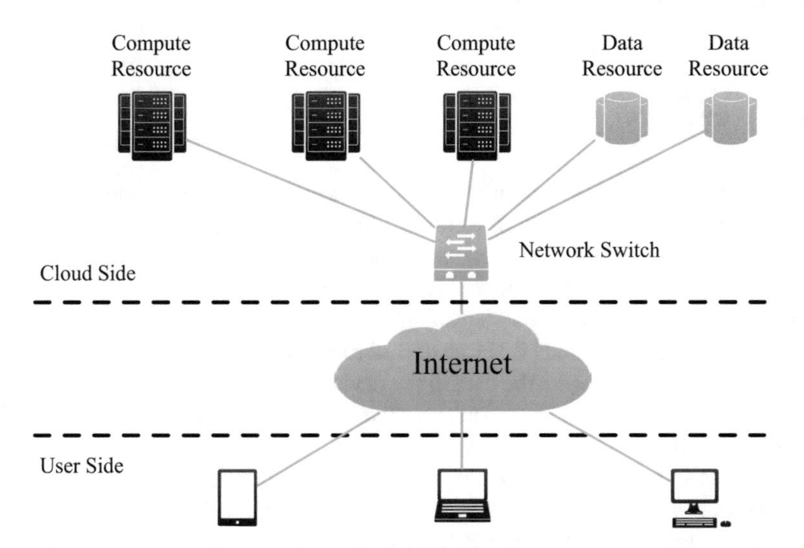

Fig. 1.4 Cloud built on top of datacenter. Nodes in datacenter are often connected through high speed network. Users access cloud service through Internet

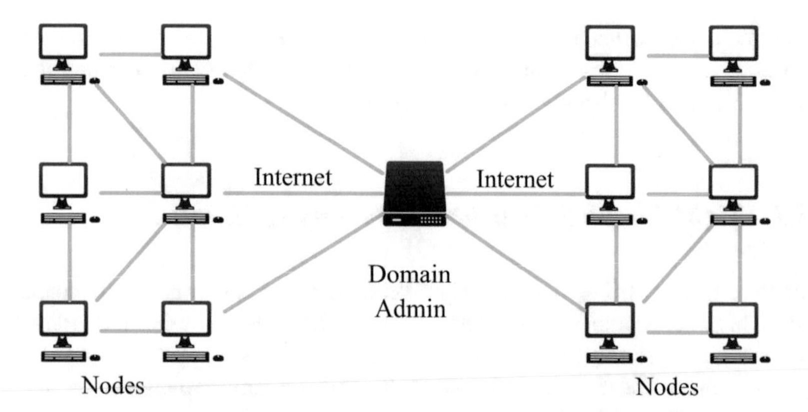

Fig. 1.5 Grid computing architecture. In this figure, the domain admin manages compute resource and data resource in the grid. The nodes are loosely coupled and connected with Internet

representative example of loose-coupled distributed memory parallel architecture. Figure 1.5 shows an example of grid architecture. As shown in this figure, each grid domain has a *domain admin* that manages compute resource and data resource in the domain. Especially, Nodes in a grid tend to be more heterogeneous and geographically dispersed (thus not physically coupled).

Grid are originally proposed for resource sharing between individual computer owners through Internet. In Grid, the resource providers are also the resource consumers. The key research point in Grid is how to organize and manage the geographically dispersed nodes into dynamic virtualized resources, so that these nodes can work together on large scale tasks.

1.4 Accelerator

Besides CPU-based parallel architecture, more and more parallel accelerators, such as Nvidia GPGPU [9] and Intel Xeon Phi [6], are proposed to speed up program execution. Accelerators are often connected with the host machine through PCIe bus. Generally speaking, when a user decides to process a program on an accelerator, the program is processed in three steps. Before an accelerator is able to process an program, the required data is first transferred to the device memory (the memory of the accelerator). Then, using the data in the device memory, accelerator executes the program. After the data is processed, accelerator returns the result to the host CPU through PCIe bus.

In this section, as the example, we introduce two widely-used accelerator, Nvidia GPGPU that executes an application in Single-Instruction-Multiple-Data (SIMD) pattern, and Intel Xeon Phi that executes an application in Multiple-Instruction-Multiple-Data (MIMD) pattern.

1.4.1 GPGPU

GPGPU is a kind of many-core architecture processor which is vastly different from CPU both in programming interface and performance characteristics. CPU can synchronize with GPU via the driver but this operation suffers from high overhead.

GPU is famous for its parallel processing ability. If the algorithm requires little to none communication between threads or these communications have good space locality, GPU can offer a significant speedup compared with CPU. The GPU is not good at executing programs with many branches and communications, some unoptimized algorithm may even be slower on GPU than CPU.

The performance of GPU varies for different sizes of workload. GPU needs a large number of running threads to hide the latency of memory access. The number of running threads is related to the workload that GPU receives. If the workload is too small, GPU will not be able to reach its full performance.

Figure 1.6 gives an example of GPGPU architecture. As shown in the figure, a GPGPU consists of multiple stream multiprocessors (SMs) and a global device memory. Each SM is a SIMD processing element. In old version of GPGPU, a GPGPU is only able to process a single kernel at a time. This constraint may result in the poor utilization of GPGPU because the workload of a kernel may not able to fully utilize all the SMs. In order to eliminate this problem, in emerging GPGPU, such as Nvidia K40, P100 etc., GPGPU starts to support concurrent kernel execution that allows multiple kernels to run on a GPGPU concurrently.

A requirement of utilizing GPGPU to speed up the execution of a parallel application, programmers need to re-write their programs using CUDA [8] or OpenCL [11]. A multi-threading application developed for CPU is not able to run on GPGPU directly.

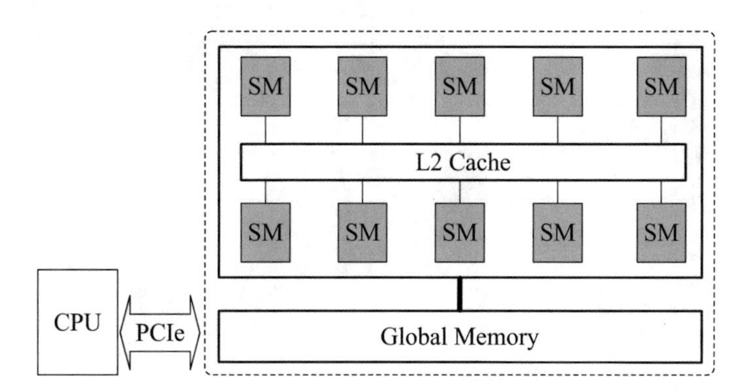

Fig. 1.6 An example GPGPU architecture. In this figure, each streaming multiprocessor (SM) is a SIMD processing element

1.4.2 Intel Xeon Phi

In order to relieve the burden of re-writing the program, Intel has proposed and released Xeon Phi accelerator. Similar to GPU, the first generation of Xeon Phi is connected with host CPU through PCIe bus. The second generation of Xeon Phi, codenamed Knights Landing was announced in 2013 [10]. The second generation Xeon Phi could be used as a standalone CPU, not just as an add-in accelerator. Multi-thread applications created for CPU can run on Intel Xeon Phi accelerator directly, although the performance is not guaranteed.

As an example, Fig. 1.7 shows the Xeon Phi (codenamed Knights Landing) architecture (denoted by *KNL* for short). As shown in the figure, a KNL processor consists of 36 tiles connected by 2D mesh interconnect, 16 GB on-chip high bandwidth MCDRAM, and the corresponding controllers. Each tile consists of two cores and 1MB L2 cache shared by the two cores; each core has two VPUs. In this case, a KNL processor has up to $36 \times 2 = 72$ hardware cores.

KNL introduces the new Advanced Vector Extensions instruction set, AVX-512, which provides 512-bit-wide vector instructions and more vector registers.3 In addition, it continues to support all legacy x86 instructions, making it completely binary-compatible with prior Intel processors.

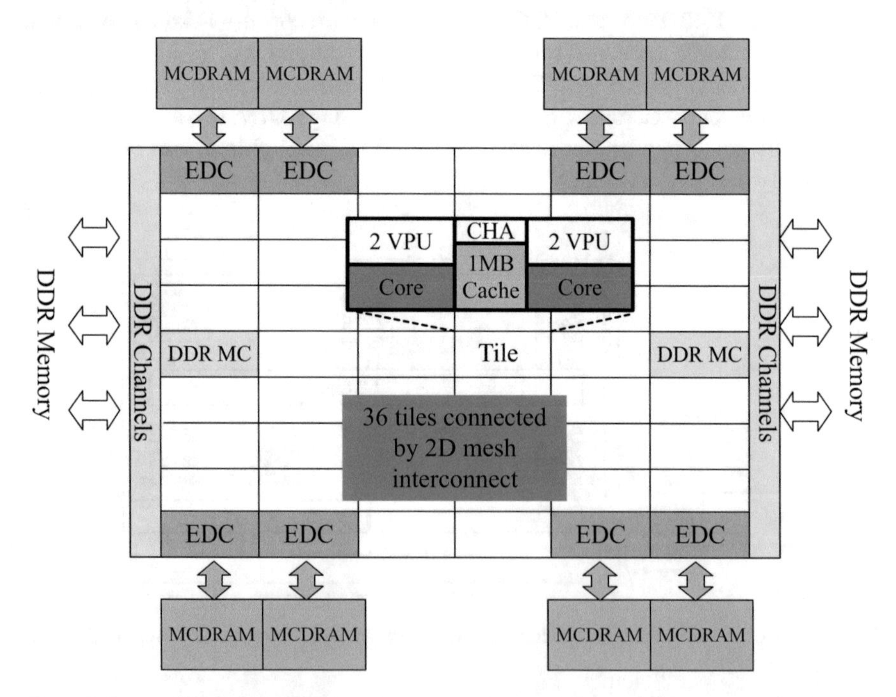

Fig. 1.7 The architecture of Xeon Phi (codenamed Knights Landing) [10]. In the figure, DDR MC is the DDR memory controller, MCDRAM is multichannel DRAM, EDC is the MCDRAM controller, and VPU is vector processing unit

KNL introduces a new 2D, cache-coherent mesh interconnect that connects the tiles, memory controllers, I/O controllers, and other agents on the chip. The mesh interconnect provides the high-bandwidth pathways necessary to deliver the huge amount of memory bandwidth provisioned on the chip to the different parts of the chip. The mesh supports the MESIF (modified, exclusive, shared, invalid, forward) cache-coherent protocol. It employs a distributed tag directory to keep the L2 caches in all tiles coherent with each other. Each tile contains a caching/home agent that holds a portion of the distributed tag directory and also serves as a connection point between the tile and the mesh. KNL has two types of memory: multichannel DRAM (MCDRAM) and double data rate (DDR) memory. MCDRAM is the 16-GB high-bandwidth memory comprising eight devices (2 GB each) integrated on-package and connected to the KNL die via a proprietary on-package I/O. All eight MCDRAM devices together provide an aggregate Stream triad benchmark bandwidth of more than 450 GB per second (GBps). KNL has six DDR4 channels running up to 2,400 MHz, with three channels on each of two memory controllers, providing an aggregate bandwidth of more than 90 GBps. Each channel can support at most one memory DIMM. The total DDR memory capacity supported is up to 384 GB. The two types of memory are presented to users in three memory modes: cache mode, in which MCDRAM is a cache for DDR; flat mode, in which MCDRAM is treated like standard memory in the same address space as DDR; and hybrid mode, in which a portion of MCDRAM is cache and the remainder is flat.

KNL supports a total of 36 lanes of PCIe Gen-3 for I/O, split into two x16 lanes and one x4 lane. It also has four lanes of proprietary Direct Media Interface to connect to the Southbridge chip, just like Intel Xeon processors. The Southbridge chip provides support for legacy features necessary for a self-booting system.

1.5 Heterogeneous Parallel Architecture

Besides the architecture introduced above, there are many other parallel architectures, such as heterogeneous parallel architecture. For instance, in emerging large-scale clusters, accelerators are often integrated with traditional shared memory parallel architecture for higher computational ability. Figure 1.8 gives several heterogenous architectures that can be used in real system.

1.6 Chapter Highlights

In this chapter, we introduced emerging popular parallel architectures, including shared-memory parallel architecture, distributed memory parallel architecture, and various parallel accelerators.

Since different architectures often have totally different characters, there is no universal task scheduling policy that can work perfectly for all parallel architectures.

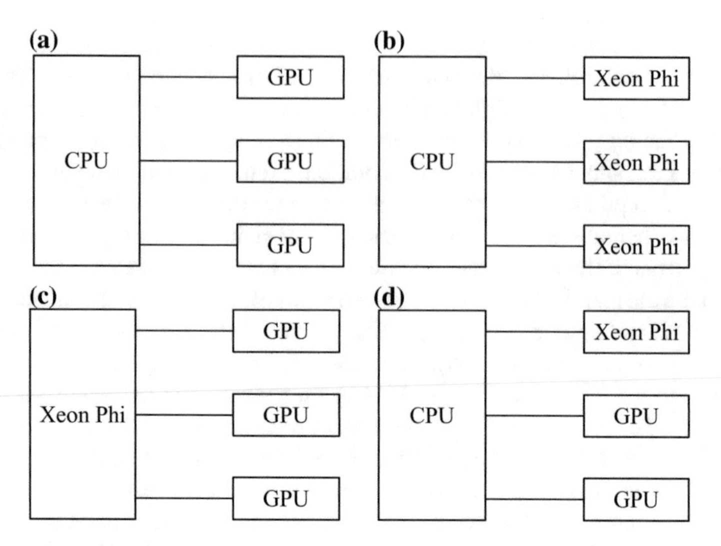

Fig. 1.8 Example of heterogeneous parallel architectures

Therefore, as parallel architectures become increase in complexity, dynamic task scheduling techniques are required to be optimized to accommodate their specific features. In the rest of this book, we introduce the recent progress in task scheduling for various parallel architectures.

References

1. S. Balakrishnan, R. Rajwar, M. Upton, and K. Lai. The impact of performance asymmetry in emerging multicore architectures. In *the 32nd Annual International Symposium on Computer Architecture*, pp. 506–517. IEEE (2005).
2. N. Chitlur, G. Srinivasa, S. Hahn, P. K. Gupta, D. Reddy, D. Koufaty, P. Brett, A. Prabhakaran, L. Zhao, N. Ijih, et al. Quickia: Exploring heterogeneous architectures on real prototypes. In *IEEE International Symposium on High-Performance Comp Architecture*, pp. 1–8. IEEE (2012).
3. H. T. Consortium. Hypertransport i/o link specification, revision 3.10c edition, 2010.
4. I. Foster, Y. Zhao, I. Raicu, and S. Lu. Cloud computing and grid computing 360-degree compared. In *Grid Computing Environments Workshop, 2008. GCE'08*, pp. 1–10. IEEE (2008).
5. P. Greenhalgh. Big. little processing with arm cortex-a15 & cortex-a7. *ARM White paper*, pp. 1–8 (2011).
6. Intel. Intel xeon phi processor (2017). https://www.intel.com/content/www/us/en/products/processors/xeon-phi/xeon-phi-processors.html.
7. Intel. Introduction to the intel quickpath interconnect. *White Paper* (2009).
8. C. Nvidia. Cuda programming guide (2010).
9. Nvidia. What is gpu-accelerated computing? (2017). http://www.nvidia.com/object/what-is-gpu-computing.html.
10. A. Sodani. Knights landing (knl): 2nd generation intel® xeon phi processor. In *Hot Chips 27 Symposium (HCS), 2015 IEEE*, pp. 1–24. IEEE (2015).
11. J. E. Stone, D. Gohara, and G. Shi. Opencl: A parallel programming standard for heterogeneous computing systems. Computing in science & engineering, 12(3):66–73, 2010.

Chapter 2
Conventional Task Scheduling Policies

Abstract In this chapter, a survey on emerging task scheduling policies will be presented respectively. There are generally two categories of task scheduling policies: manual task scheduling policies and automatic task scheduling policies. As automatic task scheduling policies are more adaptive, we further introduce several widely-used parallel programming environments that adopt automatic task scheduling policies, such as Hadoop, Spark, Cilk, X10 etc. In the last part of this chapter, we analyze the drawbacks of existing task scheduling policies.

In contrast to the quickly development of the hardware for parallel architectures, many softwares are still not effectively parallelized and thus cannot fully utilize the powerful computational ability of parallel architectures. The requirement to utilize hardware efficiently motivates the development of parallel programming environments and parallel task scheduling policies.

There are generally two categories of task scheduling policies are used in emerging popular parallel programming environments: *manual task scheduling policies* and *automatic task scheduling policies*. In manual task scheduling policies, programmers need to explicitly schedule tasks to processing elements (e.g., nodes, cores). The most popular programming environments that use manual task scheduling policies include MPI [1] and Pthreads [2]. The manual assignment of tasks is often burdensome for developing parallel applications.

In automatic task scheduling policies, parallel programs can dynamically generate tasks at runtime, and these tasks can be scheduled between the processing elements automatically. Nowadays, most well-known programming environments, such as MIT Cilk [3], Cilk++ [4], TBB [5], Java's fork-join framework [6], X10 [7], and OpenMP [8], use automatic task scheduling policies.

When a parallel program is scheduled to run on a parallel architecture, the tasks of the program are executed concurrently when the dependency between tasks are satisfied. The task scheduling policies are used to assign the parallel tasks to run on the limited numbers of processing elements. If the workload of different processing elements is not balanced, the processing element carrying the heaviest workload would

© Springer Nature Singapore Pte Ltd. 2017

Q. Chen and M. Guo, *Task Scheduling for Multi-core and Parallel Architectures*,
https://doi.org/10.1007/978-981-10-6238-4_2

degrade the performance of the whole parallel program. Therefore, task scheduling policy has a tremendous impact on the performance and energy efficiency of parallel system. Through optimizing task scheduling policy, we can greatly improve the performance of parallel programs without rewriting the parallel programs. In addition, because the automatic task scheduling policies can relieve the burden of parallelization and task assignment, they have become one edge-cutting research direction for both academia and industrial world.

2.1 Manual Task Scheduling Policies

Conventional parallel programming environments normally adopt manual task scheduling policies. In these programming environments, such as MPI and Pthreads, programmers need to manually balance the workload between threads/processes for the good performance.

2.1.1 Message Passing

Message Passing Interface (*MPI*) is a message-passing application programmer interface, together with protocol and semantic specifications for how its features must behave in any implementation. There are many efficient implementations of MPI, and one of the most well-known implementation is MPICH2 [1]. By using MPI together with C, C++ or Fortran, programmers can create parallel programs.

The purpose of MPI is to provide a compatible and effective message passing protocol for widely-used message passing programs. It is the first standardized and portable message-passing system. An MPI program can run on any parallel computers (CPU-based) without any modification. When MPI is first proposed, it targets distributed memory architecture. With the quickly development of computer architecture, MPI has already supports shared memory architecture. MPI has the following main characters.

- Although the programming model of MPI is proposed for distributed memory model, the execution of MPI programs does not rely on the low level hardware architecture.
- The parallelism is explicitly defined in MPI. Programmers have to identify the potential parallelism in their programs, and explicitly define the parallelism using interfaces provided by MPI library.

2.1.2 Multi-threading

On shared memory architecture, we can use threads to create parallel programs. In the beginning, all the hardware vendors implemented their own thread techniques respectively. Due to the diversity of thread techniques, it is challenging to develop portable and compatible parallel programs that work on different hardware. To this end, for Unix operating system, a standardized C language threads programming interface has been specified by the IEEE POSIX 1003.1c standard. Implementations that adhere to this standard are referred to as POSIX threads (Pthreads) [2].

By introducing the multi-threading method, programmers can create the parallel programs that make full use of the multiple cores. Compared with sequential model, the multi-threading programming model has the following advantages.

- It fully explores the parallelism in program and speed up the program execution.
- It supports Asynchronous I/O. When sequential program has to waiting for slow I/O operations, multi-threading program can execute other instructions instead.

2.2 Automatic Task Scheduling Policies

Compared with manual task scheduling, automatic task scheduling policies are more user-friendly. In this book, we will discuss how to design and implement efficient automatic task scheduling policies for various emerging parallel architectures. Parallel programs generally are expressed by *data parallelism* and *task parallelism*. Data parallelism is achieved when each processor performs the same task on different pieces of distributed data. And, task parallelism is achieved when each processing element executes a different thread (or process) on the same or different data. The threads may execute the same or different instructions.

2.2.1 Task Scheduling Policies for Data Parallelism

MapReduce is one of the most popular programming models that is used to express programs in data parallelism. MapReduce is not only a programming model, but also a task scheduling model. In MapReduce programming model, programmers can create *Map tasks* that process key/value pairs and generate intermediate data, and *Reduce tasks* that shuffle on the intermediate data to generate final results. The scheduling model of MapReduce [9, 10] is first proposed by Google and used in Cloud computing. Besides Google, MapReduce has also been extended to improve the performance of applications that can be expressed with high data parallelism [9–13].

MapReduce is suitable for the distributed processing of large data sets across clusters of computers. In MapReduce, programmers define the data processing pro-

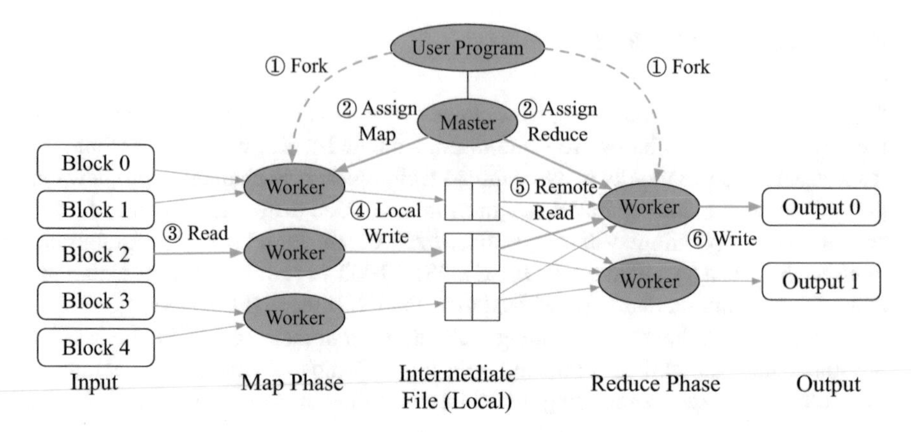

Fig. 2.1 The execution step of a MapReduce program

cedure in Map function, and define the data shuffle procedure on intermediate data in Reduce function. Programmers only need to define Map function and Reduce function to create distributed parallel programs. When a MapReduce program starts to run on a cluster/datacenter, programmers do not need to pay attention to the data splitting, allocating and scheduling. Map tasks are Reduce tasks are automatically managed by the MapReduce runtime system (e.g., Hadoop runtime system). In addition, the MapReduce runtime system is also responsible for fault-tolerance and communication managements.

Figure 2.1 shows the execution step of a MapReduce program. As shown in this figure, when a MapReduce program is invoked to run on a cluster/datacenter, it is executed in the following steps. (1) the program forks workers on distributed nodes for processing map tasks and reduce tasks. (2) the master worker of the program assign map tasks and reduce tasks to the workers housed by different nodes. (3) workers start to execute map tasks, and read corresponding data. Different workers read different data blocks (4), when a worker completes a map task, the intermediate data is written to local node to avoid data transfer through network. (5) the reduce workers then read the output of map tasks from all the nodes remotely. (6) after reduce tasks complete, the reduce workers output the final results.

It is worth noting that MapReduce uses data parallelism to speed up big data processing. In MapReduce, different Map tasks execute the same instructions but operating on different data blocks concurrently. There are low dependency between different map tasks, as well as different reduce tasks.

2.2.2 Task Scheduling Policies for Task Parallelism

Besides data parallelism, task parallelism is another widely used methodology to create parallel programs. Different from programs using data parallelism, different tasks execute different instructions in parallel programs with task parallelism. In

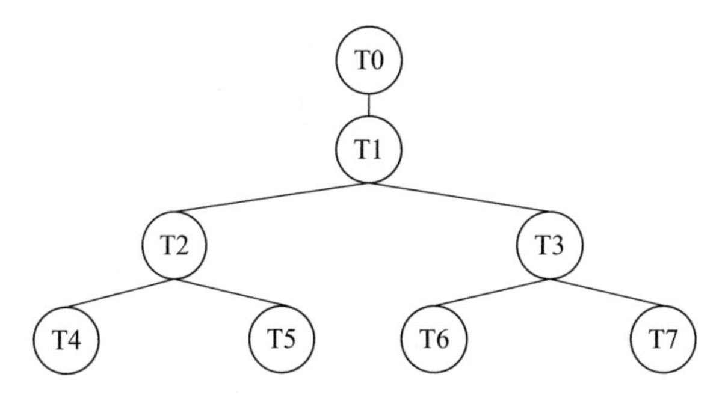

Fig. 2.2 An example of task graph (Directed Acyclic Graph, DAG)

programming environments with task parallelism, the execution of a parallel program can be represented by a task graph, which is a Directed Acyclic Graph (DAG) $G = (V, E)$, where V is a set of nodes, and E is a set of directed edges [14]. A node n_i in a DAG represents a task (i.e., a set of instructions) that must be executed sequentially without preemption. The edges in a DAG, denoted by (n_j, n_k), correspond to the dependence relationship among the nodes.

Figure 2.2 shows an example of task graph, where T_0, ..., T_7 represent tasks, and the directed edges between tasks show the dependency relationship between tasks. For example, task T_4 depends on task T_2, and only after T_2 complete, task T_4 can start to run. Task scheduling policies schedule the tasks to multiple processing elements and guarantee the dependency relationship between tasks is satisfied.

For easy of description, a processing element is called a *worker*. Therefore, the execution of a parallel program can be viewed as the parallel traversal of its task graph. *Work-sharing* [8] and *work-stealing* [15] are the two most famous task scheduling strategies for programs expressed as task parallelism.

2.2.2.1 Work-Sharing

Figure 2.3 presents the task scheduling in work-sharing policy. As shown in Fig. 2.3, in work-sharing, newly generated tasks are pushed into a centralized task pool that stores all the unexecuted tasks. When a worker is free, it tries to lock the central task pool. Once the worker successfully locks the central task pool, the worker pops a task from the pool, releases the lock on the pool, and starts to execute the newly obtained task. Because the central task pool is locked when a new task is generated and pushed into the pool, and when a worker pops tasks from the pool, work-sharing often suffers from severe lock contention. Especially, when the number of workers increases, the severe lock contention would seriously degrade the system performance. The latest OpenMP [8, 16] uses work-sharing to schedule parallel tasks.

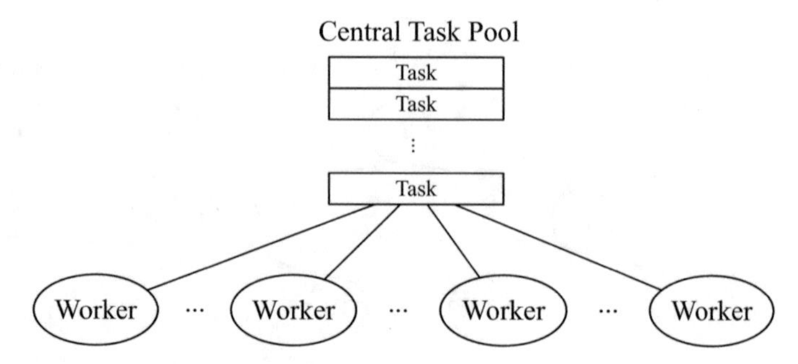

Fig. 2.3 Work-sharing task scheduling policy

2.2.2.2 Work-Stealing

In order to relieve the severe lock contention in work-sharing, researchers have proposed work-stealing policy for task scheduling. Work-stealing policy uses distributed task pool. In work-stealing policy, every worker has its own task pool. Figure 2.4 presents the task scheduling in work-stealing policy. Many programming environments and scheduling systems, such as MIT Cilk [17], TBB [5] and X10 [7] uses work-stealing policy to schedule tasks.

Most often each worker pushes tasks to and pops tasks from its own task pool without locking. Only when a worker's task pool is empty, it tries to steal tasks from other workers with locking. Since there are multiple task pools for stealing, the lock contention is much lower than work-sharing even at task steals. Therefore, work-stealing performs better than work-sharing as the number of workers increases.

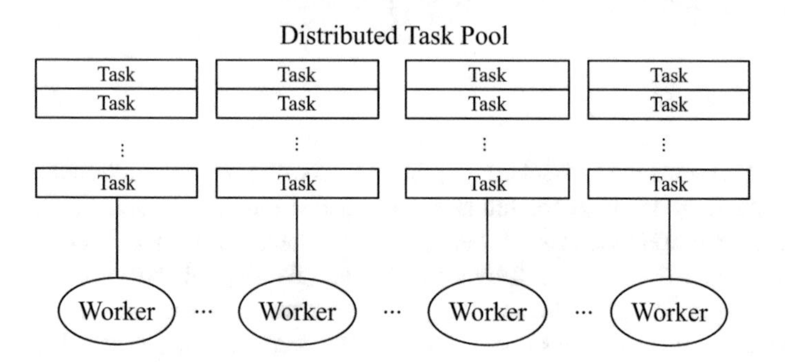

Fig. 2.4 Work-stealing task scheduling policy

2.3 Parallel Programming Environments

There are a large number of parallel programming environments have been proposed for expressing data parallelism and task parallelism. In Sects. 2.3.1 and 2.3.2, we introduce emerging widely-used parallel programming environments for data parallelism and task parallelism respectively.

2.3.1 Programming Environments for Data Parallelism

Apache Hadoop [18], Apach Spark [19], Apache Storm [20] (Heron [21]) are the most well-known parallel programming environments and task scheduling systems for expressing data parallelism. They have already been used in a large amount of real-world large scale clusters/datacenters.

2.3.1.1 Apache Hadoop

Apache Hadoop [18] is the most popular open-source implementation of MapReduce programming model. The Apache Hadoop project develops open-source software for reliable, scalable, distributed computing.

The Apache Hadoop software library is a framework that allows for the distributed processing of large data sets across clusters of computers using simple programming models. It is designed to scale up from single servers to thousands of machines, each offering local computation and storage. Rather than rely on hardware to deliver high-availability, the library itself is designed to detect and handle failures at the application layer, so delivering a highly-available service on top of a cluster of computers, each of which may be prone to failures.

The Apache Hadoop project includes these modules:

- **Hadoop Common**: The common utilities that support the other Hadoop modules.
- **Hadoop Distributed File System (HDFS)**: A distributed file system that provides high-throughput access to application data.
- **Hadoop Yarn**: A framework for job scheduling and cluster resource management.
- **Hadoop MapReduce**: A YARN-based system for parallel processing of large data sets. Hadoop MapReduce is a software framework for easily writing applications which process vast amounts of data (multi-terabyte data-sets) in-parallel on large clusters (thousands of nodes) of commodity hardware in a reliable, fault-tolerant manner.

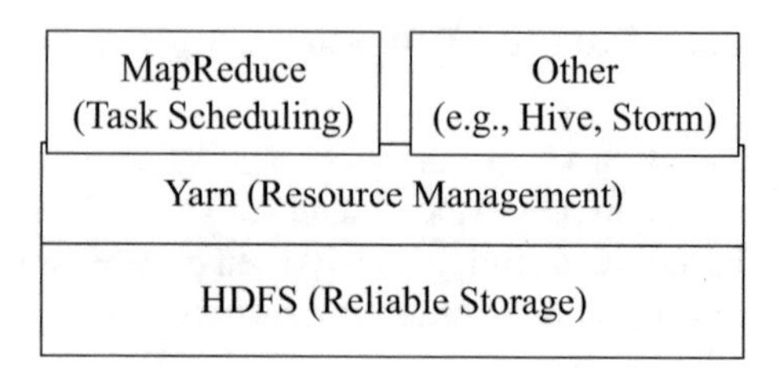

Fig. 2.5 Components in Apache Hadoop

Figure 2.5 shows the components in the latest version of Apache Hadoop (version 2.x). It is worth noting that there are many other systems (e.g., Apache Hive, Apache Storm) are implemented on top of the Hadoop Yarn resource manager.

2.3.1.2 Apache Spark

Apache Spark [19] is a fast and general-purpose cluster computing system. It provides high-level APIs in Java, Scala, Python and R, and an optimized engine that supports general execution graphs. It also supports a rich set of higher-level tools including Spark SQL for SQL and structured data processing, MLlib for machine learning, GraphX for graph processing, and Spark Streaming. Figure 2.6 shows the main components in Apache Spark.

At a high level, every Spark application consists of a driver program that runs the users main function and executes various parallel operations on a cluster. The main abstraction Spark provides is a resilient distributed dataset (RDD), which is a collection of elements partitioned across the nodes of the cluster that can be operated on in parallel. RDDs are created by starting with a file in the Hadoop file system (or any other Hadoop-supported file system), or an existing Scala collection in the driver program, and transforming it. Users may also ask Spark to persist an RDD in memory, allowing it to be reused efficiently across parallel operations. Finally, RDDs automatically recover from node failures.

A second abstraction in Spark is shared variables that can be used in parallel operations. By default, when Spark runs a function in parallel as a set of tasks on different nodes, it ships a copy of each variable used in the function to each task. Sometimes, a variable needs to be shared across tasks, or between tasks and the driver program. Spark supports two types of shared variables: broadcast variables, which

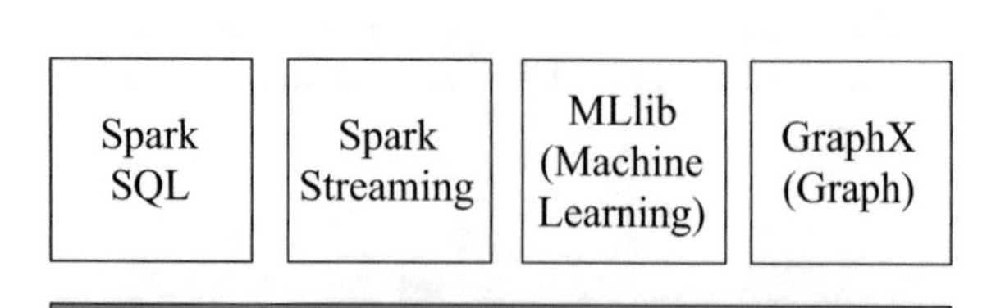

Fig. 2.6 Components in Apache Spark [19]

can be used to cache a value in memory on all nodes, and accumulators, which are variables that are only added to, such as counters and sums.

2.3.1.3 Apache Storm

Apache Storm [20] is a free and open source distributed realtime computation system. Storm makes it easy to reliably process unbounded streams of data, doing for realtime processing what Hadoop did for batch processing. Storm is simple, can be used with any programming language.

Storm has many use cases: realtime analytics, online machine learning, continuous computation, distributed RPC, ETL, and more. Storm is fast: a benchmark clocked it at over a million tuples processed per second per node. It is scalable, fault-tolerant, guarantees your data will be processed, and is easy to set up and operate.

Storm integrates with the queueing and database technologies you already use. A Storm topology consumes streams of data and processes those streams in arbitrarily complex ways, repartitioning the streams between each stage of the computation however needed.

In Storm, nodes can be classified into three categories: nimbus node (master node), zookeeper nodes, and supervisor nodes. The nimbus node is responsible for uploading jobs for execution, distributing code across all the nodes, launching workers across all the nodes, and reallocating workers as needed. The zookeeper nodes are responsible for coordinating all the supervisor nodes. The supervisor nodes actually host the workers. They communicate with nimbus node through zookeeper nodes, start and stop workers on them according to the signals from the nimbus node. Figure 2.7 shows the organization of the nodes in a storm cluster.

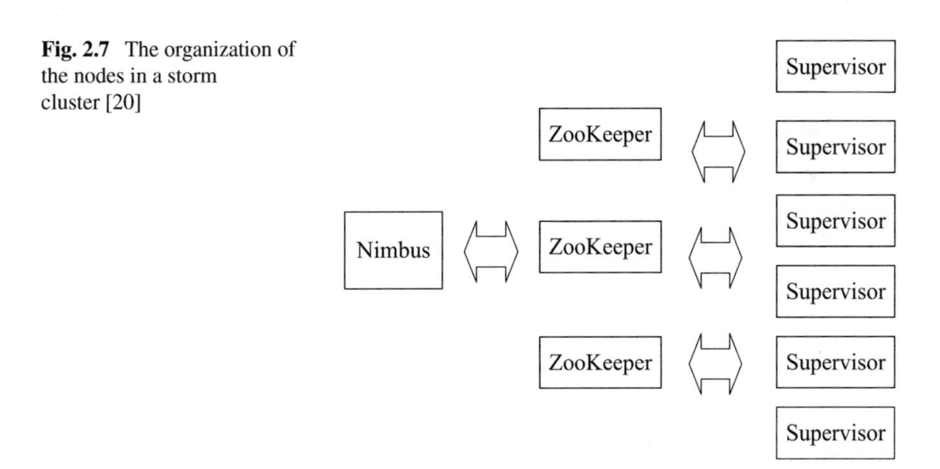

Fig. 2.7 The organization of the nodes in a storm cluster [20]

2.3.2 Programming Environments for Task Parallelism

For task parallelism, due to the good performance of work-stealing policy, academia and industry world have developed task scheduling systems using work-stealing policy. For instance, Supertech Research group at MIT developed MIT Cilk [17], IBM developed X10 [7], Intel developed Cilk Plus [4] and TBB [5]. As Cilk Plus is developed based on MIT Cilk, in this section, we introduce the widely-used MIT Cilk, TBB and X10 task scheduling system.

2.3.2.1 MIT Cilk

MIT Cilk is one of the earliest parallel programming environments that implement task-stealing [17]. It extends C with three keywords: *cilk*, *spawn* and *sync* to declare parallelism in the program. *cilk* identifies a procedure as a *Cilk procedure*, *spawn* is used to generate a child task, and *sync* waits for all the child tasks that are generated by the current task to return. Only Cilk procedures can be invoked with *spawn* as a task.

Algorithm 1 shows an simple example of MIT Cilk program. It worth noting that, if we remove the keywords *cilk*, *spawn* and *sync* in Algorithm 1, the program becomes a sequential program. MIT Cilk provides user-friendly programming interface. Using MIT Cilk, programmers can simply parallelize sequential programs by creating parallel tasks using *spawn* and inserting *sync* for synchronizing the parallel tasks. The modified programs can run on parallel architecture in parallel efficiently.

Algorithm 1 An example MIT Cilk program.

```
cilk void foo (int start, int end) {
  if(end-start < threshold) {
    more instructions;
  } else {
    int mid = (start + end) / 2;
    spawn foo (start, mid);
    spawn foo (mid, end);
    sync;
    return;
  }
}
cilk void main (int start, int end) {
  spawn foo (start, end);
  sync;
  return;
}
```

MIT Cilk consists of a compiler and a scheduler. Cilk compiler, named as *cilk2c*, is a source-to-source translator that transforms a Cilk source into a C program. *cilk2c* generates a *fast clone* and a *slow clone* for every Cilk procedure. The slow clone is executed if the task of the procedure is stolen; otherwise, the fast clone is executed instead. In addition, *cilk2c* uses a *task frame* data structure for every Cilk procedure. Once a task is generated, a task frame is created to store the information needed by the task and the scheduler. Cilk scheduler is a traditional task-stealing scheduler.

2.3.2.2 TBB

TBB (Thread Building Blocks) [5] is a set of C++ template library developed by Intel. Using the template provided by Intel TBB, programmers do not need to consider how to assign tasks to workers, and do not need to schedule the workers. There are six main modules in TBB: *Algorithm*, *Container*, *Memory Allocation*. *Synchronization*, *timing*, and *Task Scheduling*. Using the interface provided by TBB, programmers can easily define parallel tasks that can be dynamically schedule to different workers at runtime. In order to fully utilize the available resources in the parallel architecture, TBB adopts work-stealing to dynamically schedule the executable tasks. TBB provides a large amount of template functions, programmers can use these functions to create parallel tasks automatically. Algorithm 2 shows an sorting program written

Algorithm 2 An example of TBB program.

```
#include <tbb/task_scheduler_init.h>
#include <tbb/parallel_sort.h>
#include <math.h>
int main() {
  const int N = 100000;
  float a[N];
  for( int i = 0; i < N; i++ )   a[i] = sin((double)i);
  tbb::task_scheduler_init init; //Initialization
  tbb::parallel_sort(a, a + N); //Sorting
  return 0;
}
```

in TBB. Observed from the algorithm, we can find that it is easy to create parallel programs using Intel TBB. By simply replacing the sequential library call "std::sort" with "tbb::parallel_sort" provided in TBB, the sequential sorting program is updated to parallel sorting program. The task scheduling system of TBB uses work-stealing to balance the tasks generated by parallel_sort to different workers at runtime.

2.3.2.3 X10

MIT Cilk and TBB can only run on shared memory parallel architecture, and do not support distributed memory architecture. In order to solve this problem, IBM proposed the X10 [22] task scheduling system for distributed memory parallel architecture. X10 is a parallel programming and scheduling system based on Java. The programming model used in X10 is "Asynchronous, Partitioned Global Address Space, APGAS".

X10 extends the traditional programming model, and adds three keywords: *place*, *async*, and *finish*. Programmers can use *place* to define a task is created on which compute node, and use *async* to create parallel tasks, and use *finish* to create synchronization point. Functionally, the keyword *async* in X10 is similar to the keyword *spawn* in MIT Cilk; and the keyword *finish* in X10 is similar to *sync* in MIT Cilk. Meanwhile, the X10 runtime system uses work-stealing policy to schedule the tasks created with *async* to different workers running on distributed nodes.

2.4 Problems in Existing Task Scheduling Systems

However, the aforementioned task scheduling systems assume simple parallel architecture, and are not optimized against the complex parallel architectures used in real systems. Because traditional task scheduling policies lack targeted optimization, they cannot benefit from the features of the newly developed parallel architectures, which may result in poor performance.

Generally speaking, there are the following main issues that need to be resolved in order to create an effective dynamic task scheduling policy for complex parallel architectures.

- The utilization of shared cache in each socket of the MSMC architectures must be improved, to improve cache performance.
- Remote memory access on the NUMA-based shared memory system of MSMC architectures must be reduced, to reduce data access latency.
- Task distribution to the asymmetric cores in the AMC architecture must be scheduled so that the tasks are all completed at the same time.
- The workload should be balanced across heterogeneous processing elements (e.g., CPU and GPU) to achieve the best performance.
- The task scheduling policy should be optimized against big data processing, to improve data locality and relieve network congestion.
- The Quality-of-Service of user-facing application have to be guaranteed when schedule tasks to the same cluster/datacenter.

2.5 Chapter Highlights

In this chapter, we introduce widely-used task scheduling policies and the programming environments that support the corresponding scheduling policies. In the following chapters, using the random work-stealing policy proposed in MIT Cilk as the baseline, we introduce the techniques proposed to address the above problems.

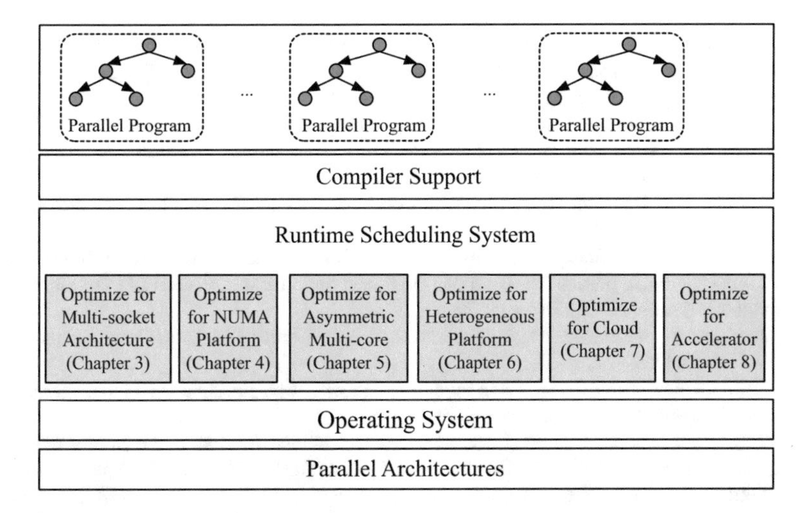

Fig. 2.8 The organization of this book. We introduce techniques proposed to improve random work-stealing for various parallel architectures

Figure 2.8 presents the organization of this book. More specifically, we first discuss task scheduling techniques for improving the performance of parallel applications on various CPU-based shared memory parallel architecture (Chaps. 3, 4 and 5). Then, we discuss task scheduling techniques for improving application performance on shared memory heterogeneous architecture consists of CPU and accelerator (Chap. 6). After that, we discuss efficient task scheduling polices for big data processing on large-scale distributed memory parallel architecture (Chap. 7). Lastly, we discuss how to schedule the co-located tasks to guarantee the QoS of high priority applications while multiple applications are executed on the same cluster/datacenter (Chap. 8).

References

1. W. Gropp. Mpich2: A new start for mpi implementations. In *Recent Advances in Parallel Virtual Machine and Message Passing Interface*, volume 2474 of *Lecture Notes in Computer Science*, pages 7–7. Springer, 2002.
2. D. Butenhof. Programming with POSIX threads. Addison-Wesley Longman Publishing Co., Inc, Boston, MA, USA, 1997.
3. R. D. Blumofe, C. F. Joerg, B. C. Kuszmaul, C. E. Leiserson, K. H. Randall, and Y. Zhou. Cilk: an efficient multithreaded runtime system. Journal of Parallel and Distributed Computing, 37(1):55–69, August 1996.
4. C. Leiserson. The Cilk++ concurrency platform. In *the 46th Annual Design Automation Conference*, pages 522–527. ACM, 2009.
5. J. Reinders. *Intel threading building blocks*. O'Reilly, 2007.
6. D. Lea. A Java fork/join framework. In *the ACM conference on Java Grande*, pages 36–43. ACM, 2000.

7. J. Lee and J. Palsberg. Featherweight X10: a core calculus for async-finish parallelism. In *Proceedings of the 15th ACM SIGPLAN symposium on Principles and Practice Of Parallel Processing*, pp. 25–36, Bangalore, India, 2010. ACM.
8. E. Ayguadé, N. Copty, A. Duran, J. Hoeflinger, Y. Lin, F. Massaioli, X. Teruel, P. Unnikrishnan, and G. Zhang. The design of OpenMP tasks. IEEE Transactions on Parallel and Distributed Systems, 20(3):404–418, 2009.
9. L. A. Barroso, J. Dean, and U. Holzle. Web search for a planet: The google cluster architecture. Micro, 23(2):22–28, 2003.
10. R. Buyya, C. Yeo, S. Venugopal, J. Broberg, and I. Brandic. Cloud computing and emerging IT platforms: vision, hype, and reality for delivering computing as the 5th utility. Future Generation Computer Systems, 25(6):599–616, 2009.
11. J. Dean and S. Ghemawat. MapReduce: a flexible data processing tool. Communications of the ACM, 53(1):72–77, 2010.
12. J. Varia. Cloud architectures. *White Paper of Amazon*, http://jineshvaria.s3.amazonaws.com/public/cloudarchitectures-varia.pdf (2008).
13. L. Vaquero, L. Rodero-Merino, J. Caceres, and M. Lindner. A break in the clouds: towards a cloud definition. ACM SIGCOMM Computer Communication Review, 39(1):50–55, 2008.
14. A. Gerasoulis and T. Yang. A comparison of clustering heuristics for scheduling directed acyclic graphs on multiprocessors. Journal of Parallel and Distributed Computing, 16(4):276–291, 1992.
15. R. D. Blumofe. *Executing Multithreaded Programs Efficiently*. Ph.D. thesis, Department of Electrical Engineering and Computer Science, Massachusetts Institute of Technology, September 1995. MIT Laboratory for Computer Science Technical Report MIT/LCS/TR-677.
16. L. Dagum and R. Menon. Openmp: an industry standard api for shared-memory programming. Computational Science & Engineering, 5(1):46–55, 1998.
17. M. Frigo, C. E. Leiserson, and K. H. Randall. The implementation of the Cilk-5 multithreaded language. In *the ACM SIGPLAN Conference on Programming Language Design and Implementation*, pp. 212–223. ACM, June 1998.
18. Hadoop. Hadoop home page. http://hadoop.apache.org/ (2011).
19. Spark. Spark home page. http://spark.apache.org (2016).
20. Storm. Storm home page. http://storm.apache.org (2016).
21. Twitter. Heron home page. https://github.com/twitter/heron (2016).
22. X10. X10: Performance and productivity at scale. http://x10-lang.org (2017).

Part II
Optimized Task Scheduling for Parallel Architectures

Chapter 3
Work-Stealing for Multi-socket Architecture

Abstract In this chapter, we discuss emerging dynamic task scheduling policies that can improve the performance of parallel applications on multi-socket architecture. In current real systems, multi-core computers often adopt a multi-socket multi-core architecture with shared caches in each socket. However, the traditional task scheduling policies (for example work-stealing) tend to pollute the shared cache and incur more cache misses. Due to the good performance of work-stealing policy, we use the traditional random work-stealing policy as the baseline in this chapter. To relieve this problem, in this chapter, we present a Cache-Aware Bi-tier work-stealing (CAB) policy. CAB improves the performance of memory-bound applications by reducing memory footprint and cache misses of tasks running inside the same CPU socket. CAB adaptively uses a task graph partitioner to divide an execution task graph into the inter-socket tier and the intra-socket tier. Tasks in the inter-socket tier are scheduled across sockets while tasks in the intra-socket tier are scheduled within the same socket. Experimental results show that CAB can significantly improve the performance of memory-bound applications compared with the traditional random work-stealing policy.

3.1 Background and Existing Problems

In *Multi-socket Multi-core* (MSMC) architecture, each CPU die is plugged into a socket and the cores in the same socket have a shared cache; the cores from different sockets can only share the main memory. However, existing task scheduling policies, such as *work-sharing* [2] and *work-stealing* [9], perform poor in the MSMC architecture.

In work-sharing, all the tasks are stored in the central task pool and a worker can execute any tasks in the pool. Therefore, from high level, tasks are randomly

---

Part of contents in this chapter has been published through IEEE Transactions on Parallel and Distributed Systems. Reprinted from Ref. [15], with permission from IEEE. Figures 3.1, 3.2, 3.6 and 3.8 in this chapter have been published through IEEE Transactions on Parallel and Distributed Systems. Reprinted from Ref. [15], with permission from IEEE

© Springer Nature Singapore Pte Ltd. 2017

Q. Chen and M. Guo, *Task Scheduling for Multi-core and Parallel Architectures*,
https://doi.org/10.1007/978-981-10-6238-4_3

scheduled to different workers in work-sharing. In work-stealing, when a worker is free and its task pool is empty, it randomly select a victim worker and steal a task from it. Because the victim worker is randomly selected, work-stealing policy schedule tasks randomly to different workers as well. This randomness causes *Task Relocation Incurred Cache Interference* (**TRICI**) problem in the MSMC architecture, which is depicted as follows.

3.1.1 The TRICI Problem

Suppose there are three independent tasks t_1, t_2 and t_3 to be executed in an MSMC architecture. Task t_1 and task t_2 share data, but they share nothing with task t_3. If t_1 and t_2 are scheduled to the workers running on cores of the same CPU socket, the shared data are loaded into the shared caches (e.g., L3 cache) only once but can be accessed by both t_1 and t_2. However, this data sharing is not respected by traditional task scheduling algorithms due to their randomness in selecting workers for the tasks. As a result, the task schedulers could move t_1 or t_2 to a worker running on a core in a different socket, where t_3 is being executed. Thus t_1 and t_2 cannot share cache and have to load data into their own caches separately.

The random task scheduling has two main weaknesses in MSMC architecture.

- First, it increases shared cache misses. Suppose t_2 is scheduled to the socket of t_3. Task t_2 cannot use the data already loaded into the caches by t_1 and have to read data from the main memory.
- Second, the random scheduling enlarges the memory footprint of the sockets. Since t_2 and t_3 share nothing but run in the same socket, the memory footprint of the socket will become larger. This increases the chance of cache misses and causes performance degradation, because t_2 may pollute the cache entries for t_3 due to conflicts or limited cache capacity.

The two weaknesses results in performance degradation problem in the MSMC architecture, which is called TRICI problem. In more detail, we use the task graph presented in Fig. 3.1 as an example to explain the TRICI problem in an MSMC architecture. In the figure, the solid lines represent the task generating relationship and the strings by the side of nodes are the identifiers of the corresponding tasks. In many parallel *divide-and-conquer* programs, neighbor tasks need to access some shared data. For example, *Heat distribution simulation algorithm* and *Successive over-relaxation algorithm* are examples of such parallel programs. Therefore, t_1 and t_2, t_3 and t_4 in Fig. 3.1 have shared data respectively.

We assume the program in Fig. 3.1 runs on a dual-socket dual-core architecture. If t_1, t_2, t_3 and t_4 are scheduled as shown in Fig. 3.2a, the shared data between t_1 and t_2 and the shared data between t_3 and t_4 is only read into the shared cache once from the main memory. Since most tasks can access the shared data in the shared cache of the socket, cache misses are reduced.

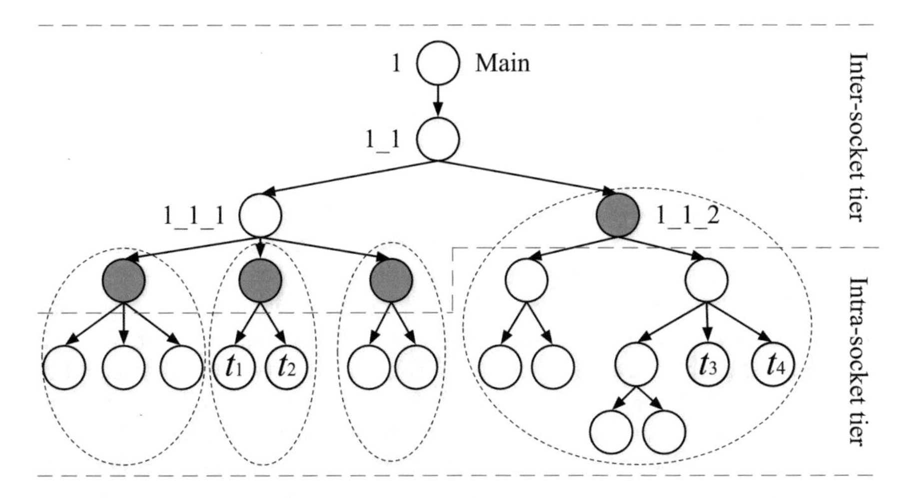

Fig. 3.1 A general task graph for divide-and-conquer programs

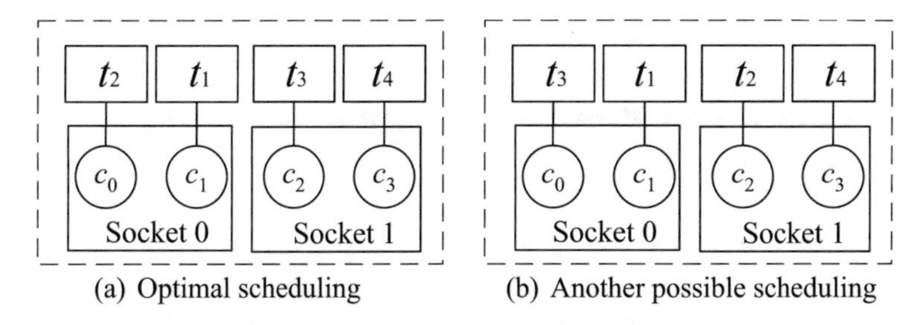

Fig. 3.2 Two possible scheduling of t_1, t_2, t_3 and t_4 on a dual-socket dual-core architecture. The first scheduling can gain performance improvement due to cache sharing and reduction of memory footprint

However, for traditional work-stealing, since it randomly chooses a victim to steal tasks, t_1, t_2, t_3 and t_4 are likely to be scheduled to the cores as shown in Fig. 3.2b. In this case, each task needs to read all its data from the main memory. This larger memory footprint leads to more compulsory cache misses. Even worse, if the memory footprint exceeds the capacity of the shared cache, the situation leads to more capacity cache misses and increases the chances of conflict cache misses. The resulted larger number of cache misses will lead to the worse performance of memory-bound applications.

3.2 Prior Solutions

Work-stealing is increasingly popular for automatic load balancing inside parallel applications. There has been a lot of research work on its adaptation and improvement [11, 23, 24, 29, 34].

Cache awareness is an interesting issue in work-stealing. In terms of theoretical work, Acar et al. [1] presented a theoretical bound on the number of cache misses for random work-stealing and implemented a locality-guided work-stealing algorithm on a single-socket SMP. Cole et al. [16] analyzed the cache misses of algorithms using random work-stealing, focusing on the effects of false sharing. Chen et al. [12] compared and analyzed cache behaviors of work-stealing and a parallel depth-first scheduler on a multi-core simulator that has shared L2 caches among cores. It was proposed to promote constructive cache sharing through controlling task granularity. However, the above studies did not take the MSMC architecture into consideration. In the following of this section, we introduce some representative work on improving the performance of parallel applications through improving data locality.

3.2.1 Scalable Locality-Aware Adaptive Work-Stealing (SLAW)

Guo et al. [22] proposed SLAW, a Scalable Locality-aware Adaptive Work-stealing scheduler. SLAW identifies and addresses two important issues that limit the scalability in current work-stealing schedulers: use of a fixed task scheduling policy, and locality-obliviousness due to randomized stealing.

There are generally two policies for work-stealing when spawning a task: *child-first* and *parent-first*. Guo et al. [21] compared the performance of the two policies in different scenarios. Under the child-first policy, the worker will execute the spawned task eagerly and leave the continuation to be stolen. Under the parent-first policy, the worker will make the spawned task available for stealing and itself will continue execution on the parent task. Child-first and parent-first policies are usually implemented with different stack and memory bounds and also exhibit performance limitations in different scenarios. The child-first policy is good for scenarios when stealing is rare. It has a provable memory bound but its implementation may overflow the stack for large irregular computations. The parent-first policy is good for scenarios when stealing is frequent. It can be implemented with low stack usage but is not space-efficient in general. In another word, a stack-based implementation of the child-first policy increases stack pressure, where as the parent-first policy can be used to reduce stack pressure (but at the expense of additional context switches).

Both child-first and parent-first policies have their strengths and are used pervasively in work-stealing schedulers. For example, MIT Cilk [8], Cilk++ [28], and Intel TBB [32] use the child-first policy, while Java's fork-join framework [25] and Task Parallel Library (TPL) [27] use the parent-first policy. However, neither child-first nor parent-first policy always perform the best. It is hard to determine a priori which policy should be used. In order to solve this problem, SLAW is designed to achieve the best of both policies, while ensuring bounded stack usage. In SLAW, tasks are generated following either the child-first policy or the parent-first policy according to the stack pressure and work-stealing conditions.

Assuming that S is the space limit (or threshold) for a worker's stack. If the spawn tree depth of a program is greater than S, then it is necessary at some point to use the parent-first policy to ensure that a worker's stack space does not exceed threshold S.

Besides the stack bound, SLAW also considers the total memory bound. The total memory bound is determined by the memory usage of both started and fresh tasks. Started tasks are those that have been executed by some processor; fresh tasks have been spawned but never executed. When only spawning under child-first policy, there will be no fresh tasks. However, under the parent-first policy, all child tasks will be created as fresh tasks and saved on the heap. In order to provide a total memory guarantee for the adaptive work-stealing scheduler: the scheduler must switch to the child-first policy when the number of fresh tasks exceeds a threshold; this ensures that the total memory used by fresh tasks are bounded.

These two conditions are enough to establish the stack and total memory bounds for the adaptive scheduling algorithm in SLAW. One thing that is important to notice is that SLAW treats stack bound as a hard bound and gives the stack condition higher priority than the fresh task condition. When the stack threshold is reached, parent-first policy will always be used to avoid stack overflow regardless of the number of fresh tasks created.

SLAW employs a runtime heuristic to select the policy if neither of the above two conditions is met. The heuristic is based on a simple estimation on the likelihood of the new spawned task being stolen. It computes the number of tasks that were stolen from the worker during the last interval. If the number of steals is greater than INT, this implies the steal rate is higher than the task creation rate. The scheduler will use the parent-first policy for the new task in the next interval to increase the rate of distributing tasks to other workers. Otherwise, the scheduler assumes the new task will not be stolen and thus uses child-first policy for the next interval to reduce the overhead of context switches.

Although SLAW uses both policies as in our solution for the TRICI problem, CAB, it does not associate the policies to the DAG tiers as in CAB (to be presented shortly). We adopt the parent-first policy to quickly generate the tasks in the inter-socket tier, but use the child-first policy to prevent the excessive task proliferation in the intra-socket tier.

3.2.2 Multi-Threaded Shepherds (MTS)

Olivier et al. [30] proposed MTS (Multi-Threaded Shepherds) to reduce cache misses in MSMC architecture. A shepherd is a group of workers affiliated with the cores in the same socket. In MTS, when all the cores in a socket are free, the head core of the socket steals a batch of tasks from other sockets.

In MTS, each socket schedules tasks depth-first locally through LIFO queue operations. An idle socket obtains more work by stealing the oldest tasks from the task queue of a busy socket. MTS implemented two probing schemes to find a victim socket: *choosing randomly* and *commencing search at the nearest socket to the thief socket*. The two schemes show similar result in MTS. In the work stealing scheduler, interruptions to busy sockets are minimized because the burden of load balancing is placed on the idle sockets. Locality is preserved because newer tasks, whose data is still hot in the processors cache, are the first to be scheduled locally and the last in line to be stolen.

The cost of work stealing operations on multi-socket multicore systems varies significantly based on the relative locations of the thief and victim, e.g., whether they are running on cores on the same chip or on different chips. Stealing between cores on different chips reduces performance by incurring higher overhead costs, additional cold cache misses, remote memory access costs, and coherence misses due to false sharing.

In order to overcome the limitations of both work stealing and shared queues, within each socket, MTS maps one worker to each core. Among workers in each socket, a shared LIFO queue provides depth-first scheduling close to serial order to exploit the shared cache. Thus, load balancing happens naturally among the workers on a chip and concurrent tasks have possible overlapping localities that can be captured in the shared cache.

Between sockets, work stealing is used to maintain load balance. Each time the sockets task queue becomes empty, only the first worker to find the queue empty steals enough tasks from the socket of another socket to supply all the workers in its socket with work. The other workers in the socket spin until the stolen work appears. Aggregate task queueing for workers within each socket reduces the need for remote stealing and decreases the number of probes required to find available work by a factor of the number of workers per socket. While a shared queue can be a performance bottleneck, the number of cores per chip is bounded, and locking operations are fast within a chip. However, MTS cannot ensure tasks executed by cores in the same socket have shared data, and thus may still suffer from the TRICI problem.

An MSMC architecture can also be viewed as an two-level hierarchical architecture, where the cores are in the lower level and the sockets are in the upper level. For distributed hierarchical architecture, prior work [31] proposed PWS (Probability Work-Stealing) and HWS (Hierarchical Work-Stealing) to reduce communications among different computers thus improve the performance of work-stealing applications.

3.2.3 Probability Work-Stealing (PWS)

In PWS [31], processors had higher probability to steal tasks from processors in the same computer. This requires a description of the hierarchy to estimate the distance between the thief and the target processor. PWS then applies the classical work-stealing algorithm with following modification. The probability to choose a target computer for steal attempts is not uniform anymore but instead proportional to the inverse of the distance between the thief and the target processor. This strategy has the advantage of increasing the data locality and of reducing the average latency of steal requests.

In the scenario of MSMC architecture, a worker has higher possibility to steal tasks from the workers running in the same socket. The downside of this technique is that the performance of the same application is not stable across different runs. This weakness of PWS makes it impractical in real-system MSMC architecture.

3.2.4 Hierarchical Work-Stealing (HWS)

Besides PWS, researchers also proposed Hierarchical Work-Stealing (HWS) [31] to improve the performance of work-stealing applications in hierarchical architecture.

By analyzing the behavior of the traditional work-stealing algorithm, the main functionality of work-stealing is to balance the workload between the thief worker and the victim worker. The main idea behind HWS is to use the same mechanism in each level. To this end, for the hierarchical MSMC architecture, HWS suggests stealing in a single steal attempt, a large amount of worker from the victim socket. It changes the choice of the stolen task as well as the target worker.

In more detail, the hierarchical platform is divided into multiple processor groups which are sets of processors connected with a fast link. For example, it could be a cluster or the set of cores in one processor. The risk of congestion between groups arises with the amount of transferred data. In order to resolve this problem, HWS choses to restrict in each group, the number of processors which can steal another group. In each group, only one processor sends remote steal requests in HWS. This processor is called a leader.

This change may however have a strong impact on load-balancing. Since the number of remote steal requests is decreased, HWS prefers remote thieves to steal a larger amount of work. Therefore, each leader donates some work to its cluster when there is not enough work, and keeps the large tasks to balance efficiently the load between leaders.

In order to distinguish tasks with a large amount of work, a limit is given by the user. The level of a task is determined according to the task graph: the level of a task is equal to the number of its parent tasks up to the root task. When new tasks are created, their level is their parent level incremented by one. In HWS, tasks are divided into *global tasks* and *local tasks*; workers are divided into *leaders* and *slaves*. Global tasks have a level below the limit and can be scheduled between different groups, while local tasks have a level higher than the limit and can only be schedule between workers in the same group.

HWS artificially limits the number of global tasks in any application. For example an application recursively dividing the work in halves like in some divide and conquer problems has 2^l global tasks where l is the chosen limit. To avoid a huge number of global tasks, l is chosen small. Global tasks are centralized on the leader.

- Leaders: they execute only global tasks. And they balance the load between groups and manage the load inside their groups.
- Slaves: they perform the classical work-stealing algorithm within their group.

In order to balance the load between groups, each leader has two stacks: the global stack for global tasks accessed by leaders, the local stack for local tasks accessed by slaves from its group. In more detail, a leader obtains a new task as follows. When the execution starts, all tasks are located on leaders stacks. Leaders which have some tasks execute them. When a task is created, the leader can choose to push the task in the local stack or in the global task. This decision depends on the task depth in the fork tree. Task pushed in the local stack by a leader is called a slave task. All tasks belonging to the fork subtree of a single slave task is called a block of tasks.

The leader provides some work to its group by pushing a local task in its own local stack. A leader detects the amount of work inside its group. If this amount is not sufficient to exploit the group processing power, the leader provides another local task to its group by executing a global task or stealing a global task. The amount of work inside the group can be estimated as proposed in the CHS algorithm [33] by sending additional messages. Another solution that can avoid such extra messages is for the leader to detect a lack of work by evaluating the number of steal attempts it receives.

HWS used a rigid boundary level to divide tasks into global tasks and local tasks which are similar to inter-socket tasks and intra-socket tasks in our solution, CAB. Different from our solution, the rigid boundary level in HWS is not adaptive.

3.2.5 CONTROLLED-PDF

The method of dividing an execution DAG into sub-DAGs for reducing cache misses was also used in other studies. For instance, Blelloch et al. [5] proposed an online scheduler, CONTROLLED-PDF, to reduce cache misses in single-socket multi-core architecture.

CONTROLLED-PDF divides the nodes of a DAG into *L2-supernodes* that contain data fit for the shared L2 cache and further divided L2-supernodes into *L1-supernodes* with data fit for the private L1 cache. By executing L1-supernodes in the same L2-supernode in parallel but executing L2-supernodes sequentially, the shared cache misses can be reduced.

Because the scheduler CONTROLLED-PDF needs users to provide space complexity function of the executed program and is only applicable to single-socket multi-core architecture, it is not practical in real-system MSMC scenario.

There are also some studies aiming at good cache performance based on other techniques. Cache-oblivious algorithms can achieve good cache performance by tuning the original algorithms carefully [7, 18]. Extended from the cache-oblivious model, a *parallel cache-oblivious* (PCO) model was proposed in [6] to account for costs on a broad range of cache hierarchies. Based on the PCO model, the authors described a scheduler to balance the cost of the cache misses across the processors. In [17], ULCC was proposed to explicitly manage and optimize last level cache usage by allocating proper cache space for different data sets of different threads based on a page-coloring technique. Although ULCC provides a good way to manage the last level cache, the management is still burdensome for programming. In contrast, CAB can improve the last level cache (L3) performance of memory-bound applications automatically.

According to the above introduction, emerging work is not able to achieve the best performance of work-stealing applications on MSMC architecture. To this end, we introduce our cache-aware bi-tier work-stealing scheme in the following part of this chapter.

3.3 Cache-Aware Bi-tier Work-Stealing

If a work-stealing scheduler can ensure tasks with shared data are scheduled to the same socket as shown in Fig. 3.2a, the shared cache misses will be minimized and the performance of memory-bound applications can be improved.

3.3.1 Solution Overview

In order to achieve the above scheduling, we analyze the execution of divide-and-conquer parallel programs shown in Fig. 3.1, and get three main observations. First, parallel tasks create child tasks recursively until the data set for each leaf task is small enough. During the procedure, only the leaf tasks physically touch the data. Second, neighbor tasks usually share some data. Lastly, if the parallel program is iterative, it often works on the same data set for a large number of iterations.

Based on the above three observations, for divide-and-conquer programs,[1] we introduce the *Cache Aware Bi-tier work-stealing* (CAB) policy that consists of a **cache aware task graph partition policy** and a **bi-tier work-stealing policy**. With the cache aware task graph partition policy, CAB can divide an task graph into *inter-socket tier* and *intro-socket tier*. With the bi-tier work-stealing scheduling policy,

[1] All the programs mentioned below are memory-bound divide-and-conquer parallel programs.

CAB schedules tasks in the inter-socket tier among sockets and schedules tasks in the intra-socket tier within socket. Because neighbor tasks in intra-socket tier often share data, the cache-aware work-stealing can better utilize the shared cache.

CAB divides an execution task graph into inter-socket tier and intra-socket tier. For instance, as shown in Fig. 3.1, the shaded tasks divide the task graph into two tiers. The shaded tasks are called *leaf inter-socket tasks*. We call ll the tasks above the leaf inter-socket tasks, including the leaf inter-socket tasks, as *inter-socket tasks*; and call the tasks in any subtree rooted with leaf inter-socket tasks as *intra-socket tasks*. All the inter-socket tasks consist of inter-socket tier and all the intra-socket tasks consist of intra-socket tier. Meanwhile, a subtree rooted with a leaf inter-socket task is called an *intra-socket subtree*. For instance, in Fig. 3.1, tasks in each ellipse consist in an intra-socket subtree. By scheduling tasks in the same intra-socket subtree within the same socket, CAB can ensure the neighbor tasks t_1 and t_2 (or t_3 and t_4) in Fig. 3.1 to be executed in the same socket.

However, it is challenging to identify the proper leaf inter-socket tasks so that tasks in the same intra-socket subtree will be able to efficiently utilize the shared cache. If an intra-socket subtree is too large (contains too much tasks), the involved data of all the tasks in the subtree could be too large to fit into the shared cache of a socket. On the other hand, if an intra-socket subtree is too small (contains too few tasks), the workload of the subtree can be too small to get better balanced among the cores of the same socket.

Therefore, in order to find the proper leaf inter-socket tasks, based on runtime information, the task graph is appropriately divided into two tiers according to the cache aware task graph partition policy. After that, inter-socket tasks and intra-socket tasks in the two tiers are scheduled differently using the bi-tier work-stealing scheduling policy. In bi-tier work-stealing, inter-socket tasks are scheduled across sockets, while intra-socket tasks are scheduled between cores in the same socket. Meanwhile, in order to avoid cache pollution, CAB guarantees that a socket cannot process tasks in multiple intra-socket subtrees concurrently. In this way, the data shared by tasks in the same intra-socket subtree is only read into shared cache once but can be accessed by all the tasks in the subtree.

3.3.2 Design Overview

In order to support cache-aware bi-tier work-stealing, we have built a runtime system for CAB as follows. As shown in Fig. 3.3, for an M-socket N-core architecture, A-CAB launches $M \times N$ workers (i.e., threads) at runtime and affiliates each worker with one individual hardware core. For convenience of presentation, we use the term *core* to mean a worker in the rest of this chapter.

In each socket, only one core is selected as the head core of the socket to look after the inter-socket task scheduling. In CAB, we choose "core 0" in each socket as the socket's head core.

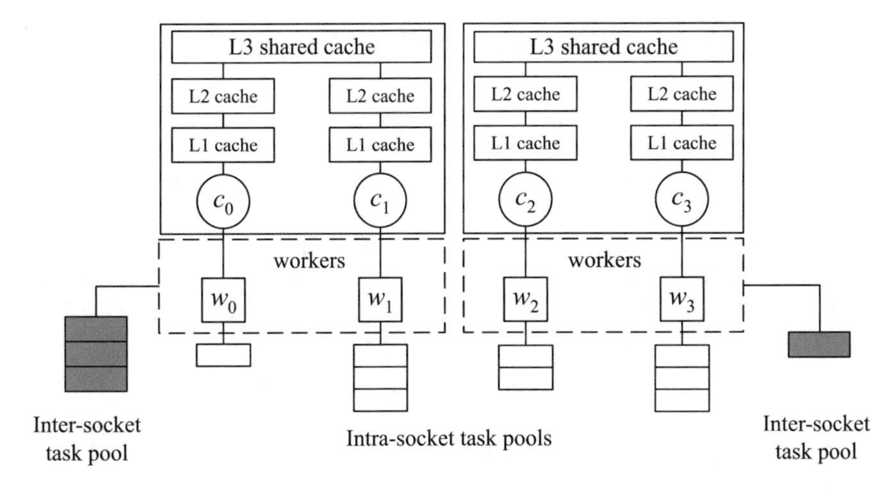

Fig. 3.3 CAB runtime environment in a dual-socket dual-core architecture. Each socket has an inter-socket task pool and each core has an intra-socket task pool

As shown in Fig. 3.3, in order to schedule inter-socket tasks and intra-socket tasks in different way, CAB creates an *inter-socket task pool* for each socket to store inter-socket tasks, and an *intra-socket task pool* for each core to store intra-socket tasks. When a worker w in any socket ρ generates a new task t, and t is an intra-socket task, it is pushed into the intra-socket task pool of w. Otherwise, if t is an inter-socket task, it is pushed into the inter-socket task pool of ρ. Similar to prior work-stealing scheduler implementations [32], in CAB, a task pool is implemented as a double-ended queue.

In cache-aware bi-tier work-stealing policy, if a worker is free, it first tries to obtain a new task from its own intra-socket task pool. If its own intra-socket task pool is empty, the worker tries to steal a task from the intra-socket task pools of the workers in the same socket. If all the intra-socket task pools in its socket is empty, the head worker of the socket tries to obtain an inter-socket task from its inter-socket task pool. If its inter-socket task pool is empty as well, the head worker tries to steal an inter-socket task from inter-socket task pools of other sockets.

In order to relieve the lock contention on inter-socket task pools, in our work-stealing policy, only the head worker is allowed to steal inter-socket tasks from the inter-socket task pools of other sockets. Meanwhile, CAB does not allow a socket to process tasks from multiple intra-socket subtree at the same time. This is because the intra-socket tasks from different intra-socket subtree would pollute the shared cache, thus increase shared cache misses and degrade the performance of parallel programs.

3.4 Cache-Aware Task Graph Partition Policy

As mentioned before, one main challenging problem in cache-aware bi-tier work-stealing is how to properly divide an task graph into inter-socket tier and intra-socket tier. Targeting this problem, in this section we introduce two task graph partition policies: *Full Tree Oriented* (FTO) partition policy, and *General Tree Oriented* (GTO) partition policy. If a task graph is a full tree, in which all the nodes have the same number of children nodes (except leaf node), FTO policy is applicable. Otherwise, GTO policy is applicable. In the following subsections, we first introduce the FTO policy, and then discuss the generalized GTO policy.

3.4.1 Full Tree Oriented Partition Policy

The full tree oriented partition policy only applies for parallel programs where the task graphs are full tree. Because most of existing parallel divide-and-conquer programs create tasks recursively, most of these programs' task graphs are full tree and FTO policies apply for them. Figure 3.4 presents a five level full binary tree task graph (except node t_0).

Because the task graph in Fig. 3.4 is a regular full binary tree, all the leaf inter-socket tasks are in the same level of the task graph. If we can find the level where the leaf inter-socket tasks are in (called "Boundary Level"), all the leaf inter-socket tasks can be identified. And, these leaf inter-socket tasks divides the task graph into inter-socket tier and intra-socket tier. For easy of description, we use L_b to represent the boundary level.

Figure 3.5 presents the processing flow of a parallel program in CAB if FTO partition policy is adopted to divide its task graph into two tiers. In FTO scheduling policy, it is challenging to identify the appropriate boundary level. As shown in

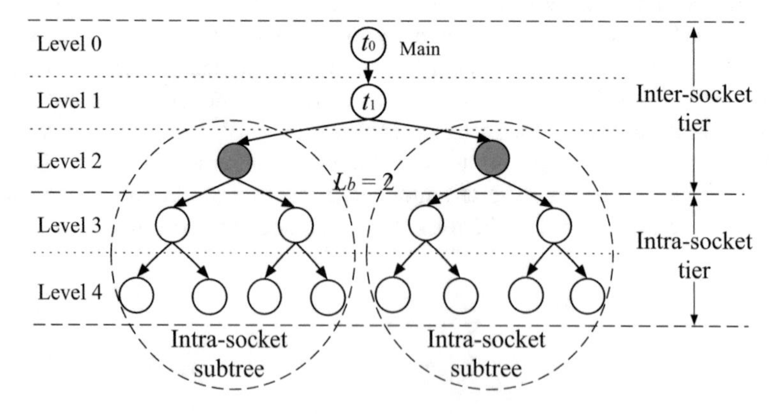

Fig. 3.4 An example of full binary tree task graph (except node t_0)

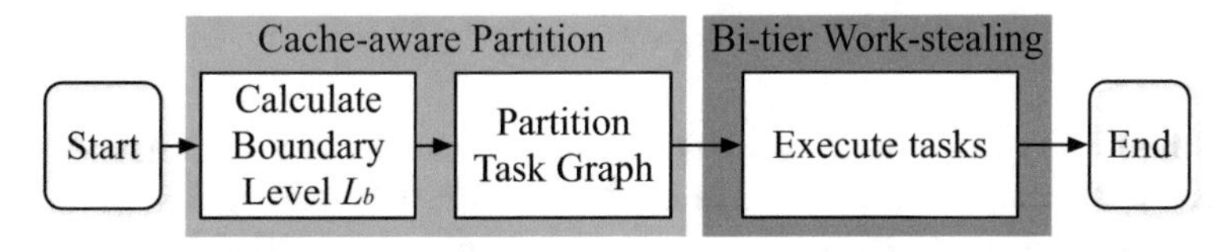

Fig. 3.5 The processing flow of a parallel program in CAB if it adopts FTO policy to divide task graphs

Fig. 3.4, if the boundary level is too high (e.g., level 1), the intra-socket tier contains too many tasks. In this scenario, the involved data of an intra-socket subtree could be too big to fit into the shared cache of a socket. On the other hand, if the boundary level is too low (e.g., level 4), the intra-socket tier contains too few tasks. In this scenario, the number of tasks in each intra-socket subtree is too small to balance the workload between cores in a socket.

Therefore, an appropriate boundary level should obey three main constraints. First, there should be enough leaf inter-socket tasks so that there are enough intra-socket subtree to fully utilize all the sockets. Second, the involved data of each intra-socket subtree should be small enough to fully fit into the shared cache of a socket. Third, each intra-socket subtree should have enough tasks, so that the workload of an intra-socket subtree can be balanced across different workers in a socket. In FTO scheduling policy, we can use three parameters: the effective input data size, the shared cache size of a socket, and the out degree of a task in the task graph, to find the appropriate bound level.

In the following discussion, we assume that the program directly generates the task of the recursive divide-and-conquer procedure in the "main" procedure, which is the case for all our benchmarks. For example, in Fig. 3.4, the main task t_0 directly spawns task t_1 that recursively spawns tasks executing itself until a cut-off point. However, if the recursive procedure is not directly generated by "main", we need either manual adjustment of the L_b value, or compiler support to adjust L_b automatically.

In the model, we suppose a parallel program that has full tree task graph run on an M-socket N-core architecture, in which each socket has shared cache of size S_c. We further suppose the effective input data size of the program is S_d, and the program divides the input data into B parts each time sub-tasks are generated, i.e., the out degree of each task in the task graph of the recursive procedure is B. In this scenario, because every task generates B tasks for the next level and this is repeated for $L_b - 1$ times until the boundary level L_b, there are B^{L_b-1} leaf inter-socket tasks in the boundary level. For instance, in Fig. 3.4, because the out degree of every task $B = 2$, and the boundary level $L_b = 2$, the boundary level (level 2) contains $2^{2-1} = 2$ leaf inter-socket tasks. In other word, this program has two intra-socket subtrees. Table 3.1 presents the parameters used in the full tree oriented partition policy for finding the appropriate boundary level L_b.

Because there are M sockets, according to the first constraint, in order to balance the workload between sockets, we have to guarantee there are at least M leaf inter-socket tasks. Therefore, the boundary level L_b need to fulfill Eq. 3.1.

$$D^{L_b-1} \geq M \tag{3.1}$$

Because every task equally divides its data set for sub-tasks, the data set size of each leaf inter-socket task can be calculated as S_d/D^{L_b-1}. In order to fulfill the second constraint (the involved data of an intra-socket subtree should be able to store in the shared cache of a socket), the constraint can be expressed with Eq. 3.2.

$$\frac{S_d}{D^{L_b-1}} \leq S_c \tag{3.2}$$

From Eqs. 3.1 and 3.2, we can deduce two conditions for selecting an appropriate value for boundary level L_b, as shown in Eq. 3.3.

$$\begin{cases} L_b \geq \log_D M + 1 \\ L_b \geq \log_D (S_d/S_c) + 1 \end{cases} \tag{3.3}$$

From Eq. 3.3, we can select any L_b that is large enough to satisfy the two inequations. But, unfortunately, if L_b is too large, the number of the intra-socket tasks generated by a leaf inter-socket task will be too small, which leads to poor load balance within a socket. Therefore, in order to satisfy the third constraint, we set L_b to be the smallest value that satisfies both inequations in Eq. 3.3, as shown in Eq. 3.4.

$$L_b = \max\{\lceil \log_D M + 1 \rceil, \lceil \log_D (S_d/S_c) + 1 \rceil\} \tag{3.4}$$

In the implementation of FTO partition policy, we can use a semi-automatic way to obtain all the parameters, and then calculate the boundary level L_b using Eq. 3.4. Parameters M and S_c are automatically acquired from "/proc/cpuinfo" by the runtime system, D can be acquired by the compiler through program analysis, but S_d should be provided through command line argument. Once the task graph is divided into two tiers, a bi-tier work-stealing scheduler is used to schedule the tasks to minimize shared cache misses.

Table 3.1 Parameters used in the full tree oriented partition policy

Parameters	Description
L_b	Boundary level
D	The out degree of a task
M	The number of sockets
S_d	The size of input data
S_c	The capacity of a socket's shared cache

3.4.2 General Tree Oriented Partition Policy

The FTO partition policy introduced in Sect. 3.4.1 has one limitation, it can only apply for divide-and-conquer programs that have full tree task graphs. If the task graph of a parallel program is not a full tree, the leaf inter-socket tasks could be in different layers. In this case, FTO partition policy is not able to appropriately divide the task graph into two tiers. For example, the FTO partition policy is not able to divide the irregular task graph in Fig. 3.1 into two tiers appropriately. In order to solve this problem, we introduce the general tree oriented partition policy that works for irregular task graphs based on compiler support and information collected dynamically at runtime.

In the *General Tree Oriented* (GTO) partition policy, in order to find the proper leaf inter-socket tasks, we need first to obtain the data size involved in each task. The GTO policy uses a *profiling-based method* and a *compiling-based method* to obtain the size of data involved in each task for iterative programs and non-iterative programs respectively. Based on the size of data involved in each task, the GTO partition policy can divide the task graph into two tiers appropriately. Figure 3.6 presents the processing flow of a parallel program in CAB if GTO partition policy is adopted to divide its task graph into two tiers.

As we present before, to partition a task graph into inter-socket tier and intra-socket tier optimally, the most challenging problem is to find the proper leaf inter-socket tasks. Once the proper leaf inter-socket tasks are identified, the task graph can be easily divided into two tiers: all the tasks above the leaf inter-socket tasks (including the leaf inter-socket tasks) belong to the inter-socket tier, and those tasks in the subtrees rooted with leaf inter-socket tasks belong to the intra-socket tier.

In the GTO partition policy, an optimal partitioning of a task graph should satisfy two constraints as well. The first constraint is that, for any intra-socket subtree ST, the involved data of all the tasks in ST is small enough to fit into the shared cache of a socket. The second constraint is that an intra-socket subtree ST should be large enough to allow a socket to have sufficient intra-socket tasks.

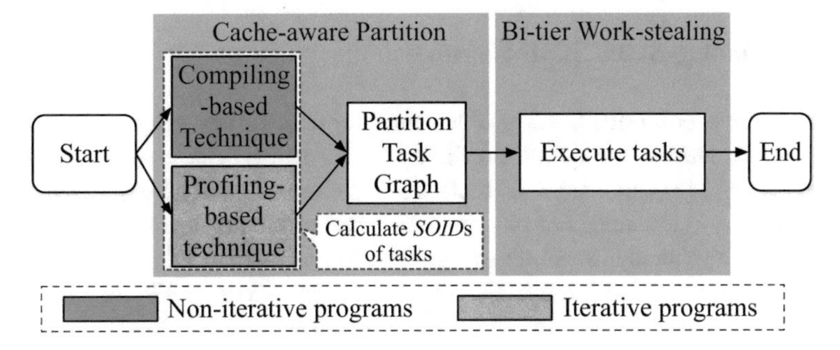

Fig. 3.6 The processing flow of a parallel program in CAB if it adopts GTO policy to divide task graphs. The SOID of a task is defined as the size of involved data of the task

To fulfill the two constraints when dividing a task graph, for any task t in the task graph, we should collect its involved data size. For convenience of description, in this chapter, we use *Size Of Involved Data* (SOID) to represent the involved data size of a task t. That is, SOID includes the data accessed by all tasks in the subtree rooted with task t. Once the SOIDs for all tasks in the task graph are known, the task graph partitioner can divide the task graph into two tiers optimally. As stated before, the GTO partition policy uses the profiling-based method to collect SOIDs of tasks for iterative programs while using the compiling-based method to collect SOIDs of tasks for non-iterative programs.

3.4.2.1 Compiling-Based SOID Calculation

In the compiling-based SOID calculation method, for any task t, we calculate its SOID using the effective input data size of the program and the branching degree of all its ancestors in the task graph. Note that, in the following calculation, we still assume that a task divides its data set into several parts evenly according to its branching degree. This assumption is true in most of existing divide-and-conquer programs.

Suppose the effective input data size of the program is S_{input}, the ancestors of task t in the task graph are t_0, t_1, ..., t_i whose branching degrees are B_0, B_1, ..., B_i accordingly. Then, the SOID of task t, denoted by D, can be calculated with Eq. 3.5.

$$D = \frac{S_{input}}{\prod_{k=0}^{i} B_k} \tag{3.5}$$

In Eq. 3.5, the branching degree of each task can be obtained by analyzing the task generating pattern in the source code through the compiler. However, the effective input data size of the application S_{input} has to be provided through a command line argument. The SOIDs of tasks are calculated in a top-down way in the compiling-based method for non-iterative programs.

3.4.2.2 Profiling-Based SOID Calculation

For an iterative program, CAB profiles the program during the first iteration of the execution. During the online profiling, the hardware Performance Monitoring Counters (PMC) [3] can be used to collect cache misses, based on which the SOIDs for all tasks are calculated. For the processors where the last level cache misses of each core can be collected separately (e.g., Intel Xeon E5-2650 V4), the performance counter event we can use is the last level shared cache accesses. On the other hand, for processors which do not support the above event (e.g. AMD Quad-core Opteron 8380), we can use the last level private data cache misses. For detailed information of the performance counter events, refer to *BIOS and Kernel Developer's Guide* of the corresponding processor. For instance, in AMD Quad-core Opteron 8380, we can use the performance counter event "07Eh" with mask of "02h" to collect the last level private data cache misses.

Though it is straightforward to collect the event statistics of the last level private data cache misses and the last level shared cache misses in modern multi-core machines like X86/X64 that support corresponding events, it is very tricky to calculate the SOIDs of the tasks based on the collected event statistics.

First, limited by the hardware PMCs, a core can only collect the cache misses of its own, but a task may have multiple child tasks executing on different cores. Therefore, it is impossible to collect the overall cache misses for a task directly.

Second, it is nontrivial to relate the private cache misses on processors where collecting per core shared cache misses is not supported, to the SOID of a task. For a task t that runs on a core c in socket ρ, if t fails to get its data from the last level private cache of c, it requests the data from the shared cache of ρ. Since c does not execute other tasks when it is executing t, the last level private cache misses of c are totally caused by t. The last level private cache misses of c can be used to approximate to the size of data accessed by t for the following reasons. Many memory-bound applications adopt data parallelism. As mentioned in our second observation in Sect. 3.1.1, only the leaf tasks physically access data. The data of leaf tasks do not have much overlapping with each other. Even when two neighbor leaf tasks have a small portion of shared data, the chances for them to be executed in the same core are small in a random work-stealing scheduler, which is adopted during the profiling stage. Therefore, the above approximation is accurate enough for us to calculate the SOIDs of all tasks.

Based on the collected last level private cache misses (or the last level shared cache accesses) of task t, its SOID is calculated as follows. If t is a leaf task, the number of last level private cache misses (or the last level shared cache accesses) of task t times the cache line size (e.g., 64 bytes in AMD Quad-core Opteron 8380) is task t's SOID. Otherwise, if t is not a leaf task, its SOID is the sum of its last level private cache misses (or last level shared cache accesses) times the cache line size plus the SOIDs of all its child tasks. Given a task t with n sub-tasks t_1, t_2, ..., t_n. Suppose M is task t's number of last level private cache misses (or last level shared cache accesses) times the size of cache line, and the SOIDs of its child tasks are D_1, D_2, ..., D_n respectively, then t's SOID, denoted by D, is calculated as in Eq. 3.6.

$$D = M + \sum_{i=1}^{n} D_i \tag{3.6}$$

Based on Eq. 3.6 and Fig. 3.7 presents an example of calculating SOIDs for all the tasks. In the figure, D_i is the SOID for leaf task t_i, but represents the size of data physically accessed by the task itself for non-leaf tasks. In fact, for many memory-bound applications, D_i for non-leaf tasks is very small, if it is not zero, since non-leaf tasks do not physically access data.

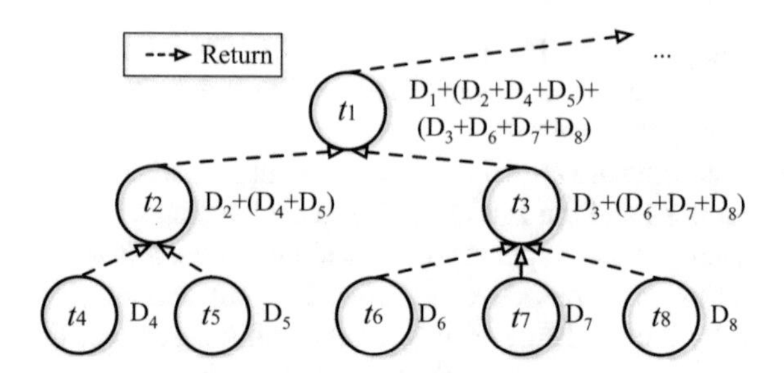

Fig. 3.7 Calculate the Size Of Involved Data (SOID) for tasks in iterative programs

As shown in Fig. 3.7, the SOID of a task is returned to its parent task when it complete. For example, in Fig. 3.7, task t_2's SOID is added to t_1's SOID when task t_1 complete. Therefore, when all the tasks in the first iteration are completed, the SOIDs of all the tasks can be calculated. The SOIDs of tasks are calculated in a bottom-up way in the profiling-based method.

3.4.2.3 Task Graph Partitioning

No matter the semi-automatic compiling-based technique or the full-automatic profiling-based technique is adopted, once the SOIDs of all the tasks are calculated, the task graph partitioner divides the task graph into inter-socket tier and intra-socket tier automatically.

In order to satisfy the constraints in Sect. 3.4.2, we can identify the leaf inter-socket tasks obeying the following rules. For a task t and its parent task t_p, let D and D_p represent SOIDs of task t and t_p respectively. In this scenario, t is a leaf inter-socket task if and only if D is smaller than the size of the shared cache and D_p is larger than the size of the shared cache. More precisely, given a task t and its parent task t_p, we can determine t's tier as follows.

- If both D_p and D are larger than the shared cache of a socket, t is an inter-socket task, as shown in Fig. 3.8a.
- If D_p is larger than the shared cache of a socket and D is smaller than the shared cache of a socket, t is a leaf inter-socket task, as shown in Fig. 3.8b.
- If both D_p and D are smaller than the shared cache of a socket, t is an intra-socket task, as shown in Fig. 3.8c.

After we determine the tiers of each and every task, the task graph has already been divided into two tiers appropriately. Based on the partitioned task graph, we use bi-tier work-stealing policy to schedule tasks in the task graph for optimizing shared cache usage.

For iterative programs, in order to identify the same task in the following iterations, during the execution of a parallel program, each task is given an identifier (a string) according to the spawning relationship between tasks. If a task t's identifier is S, then

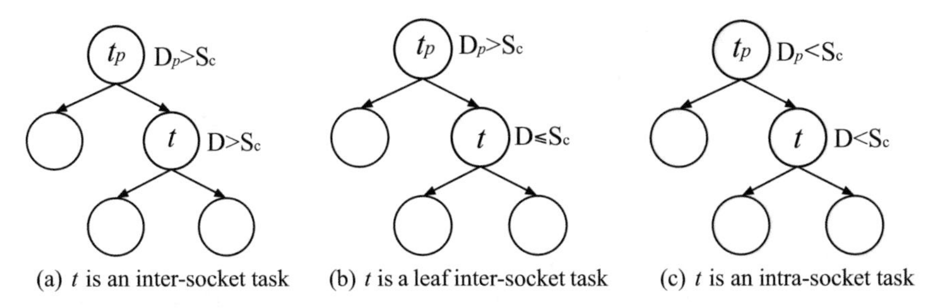

(a) t is an inter-socket task (b) t is a leaf inter-socket task (c) t is an intra-socket task

Fig. 3.8 Conditions that t is an inter-socket task, a leaf inter-socket task or an intra-socket task

its ith sub-task's identifier is S_i. For example, Fig. 3.1 shows the way of constructing identifiers for tasks. The strings beside the tasks are the identifiers in Fig. 3.1. The identifiers of all the completed tasks are saved in a hash table with their SOIDs. When a new task is spawned, CAB tries to find its identifier in the hash table. If the identifier is found, it means the first iteration has completed since a new task in the same location of the task graph has been spawned. In this case, CAB uses the bi-tier work-stealing scheduler to schedule tasks based on their tiers which are decided according to their SOIDs as shown above.

It is worth noting that, for iterative programs, all the needed information for the optimal bi-tier work-stealing can be obtained automatically by the runtime system. In this way, CAB can automatically improve the performance of iterative programs without any human intervention. In addition, for non-iterative programs, CAB only needs users to provide the input data size of the programs while all the other information can be obtained by the compiler automatically. In this way, CAB can automatically improve the performance of non-iterative programs with slight human intervention. Compared with the Full-Tree Oriented (FTO) partition policy described in Sect. 3.4.1, the profiling-based technique and the compiling-based technique in the GTO partition policy are adaptive to irregular and unbalanced task graphs where the leaf inter-socket tasks do not have the same depth, because it determines the leaf inter-socket tasks according to the SOIDs of tasks instead of the task's depth in the task graph. In other words, the GTO task graph partition policy can be applied to partition more general task graphs.

3.5 Bi-tier Work-Stealing Scheduling Policy

After the task graph of a parallel program is divided into inter-socket tier and intra-socket tier, the following bi-tier work-stealing scheduler can schedule the tasks in two tiers in different ways accordingly. In this section, we discuss the bi-tier work-stealing task scheduling policy that can minimize shared cache misses and thus improve the performance of parallel programs. The task stealing policy is used by a free worker when it tries to obtain a new task.

3.5.1 Work Stealing Algorithm

When a parallel program starts to run, the runtime system first decides which policy can be used to schedule the program. There are three cases where different algorithms are used to schedule tasks.

- First, if the FTO policy (Sect. 3.4.1) is used to partition the task graph, each worker uses bi-tier work-stealing policy to schedule the tasks.
- Second, if the GTO policy (Sect. 3.4.2) is used instead and the parallel program is iterative, its task graph has not been divided into two tiers during the first iteration. Therefore, in the first iteration, workers adopt traditional random work-stealing policy to obtain or steal a new task. In the following iteration, each worker adopts a bi-tier work-stealing algorithm to schedule tasks so that tasks in a subtree rooted with a leaf inter-socket task are scheduled to the same socket.
- Third, if the GTO policy (Sect. 3.4.2) is used to partition task graph and the program is a non-iterative program, each worker directly adopts the bi-tier work-stealing algorithm to schedule tasks.

When a parallel program is launched, a command line argument can be used to indicate whether a program is iterative or not. Because the details of the traditional random work-stealing policy has already been introduced in Sect. 2.2.2.2, we only present the bi-tier work-stealing policy here.

In CAB runtime system, when a worker w affiliated with a core in socket ρ completes its current task and becomes free, it first tries to obtain a task from its own intra-socket task pool. If its intra-socket task pool is empty, w tries to steal a task from the intra-socket task pools of other workers in ρ. As we described in Sect. 3.3.2, in each socket, we choose one of the worker as the header worker of the socket. If the intra-socket task pools of all the workers in socket ρ are empty, the head worker of ρ tries to obtain a task from its inter-socket task pool. If its inter-socket task pool is empty as well, the head worker tries to steal an inter-socket task from other sockets.

The above task stealing algorithm should obey two constraints. First, only the head worker of each socket can steal inter-socket tasks. If all the free workers can steal inter-socket tasks, the lock contention of the inter-socket task pools is too heavy which in turn degrades the performance of parallel programs. Second, workers in the same socket are not allowed to execute tasks in different intra-socket subtrees at the same time. This policy can avoid the situation where different intra-socket subtrees pollute the shared caches with different data sets. The downside of the policy is that some workers in a socket may be idle waiting for other workers to finish their tasks. An alternative policy is to allow a socket to execute tasks from more than one intra-socket subtrees at the same time. This alternative policy can ensure most cores are busy, but different intra-socket subtrees may pollute the shared caches, which leads to more cache misses. For the memory-bound applications that CAB is targeting, the cache misses are more critical to the performance according to our experimental results. Therefore, we have adopted the first policy in CAB.

In order to fulfill the above two constraints, we use a boolean variable *busy_state* in each socket to indicate whether there is an inter-socket task running in the socket.

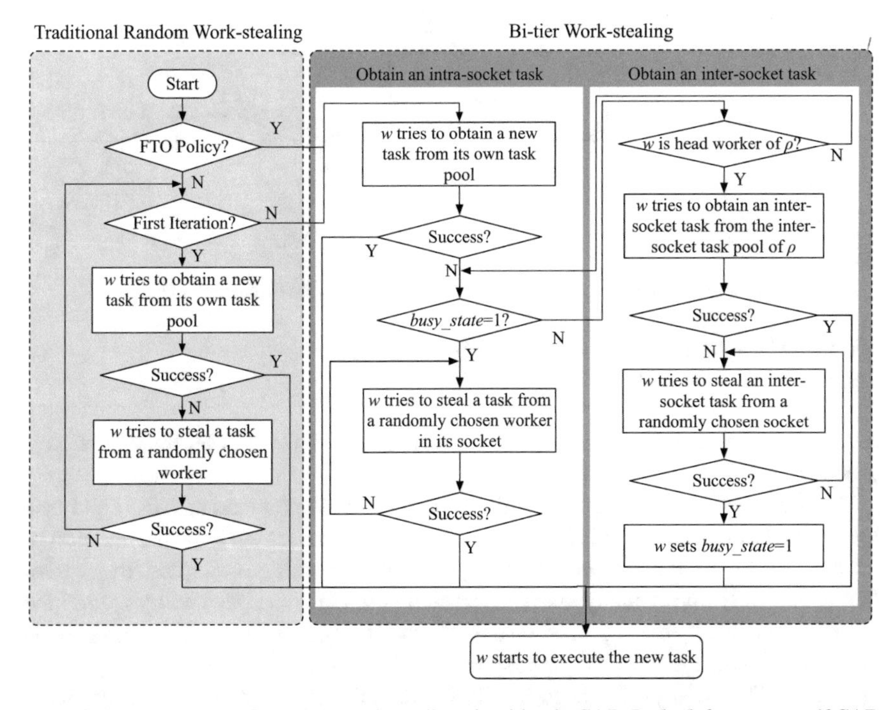

Fig. 3.9 The flow chart of the bi-tier task stealing algorithm in CAB. In the left-most part, if CAB uses GTO method to partition task graph and it is the first iteration of the program, CAB uses the traditional work-stealing to schedule the program. If it is not the first iteration of the program, in the middle part, a worker tries to obtain/steal an intra-socket task from workers in its socket. In the right-most part, the header worker in the socket tries to obtain/steal an inter-socket task

If a socket obtains or steals an inter-socket task successfully, its *busy_state* is set true (1). Once the socket finishes its inter-socket task, its *busy_state* is set false (0). Only if *busy_state* is false, should the socket obtain or steal another inter-socket task. Suppose w is a worker in a socket ρ, and w is free and trying to get a new task. Figure 3.9 shows the detailed bi-tier work-stealing algorithm used by a free worker w in socket ρ.

3.5.2 *Task Generating Algorithm*

In CAB runtime system, inter-socket tasks and intra-socket tasks can be generated in different policies. Generally speaking, two types of task-generating policies, child-first and parent-first, can be adopted for work-stealing. Figure 3.10 shows how Algorithm 1 is executed with the child-first and the parent-first policy. In the figure, the solid arrows represent the spawn operations, the dashed arrows represent the

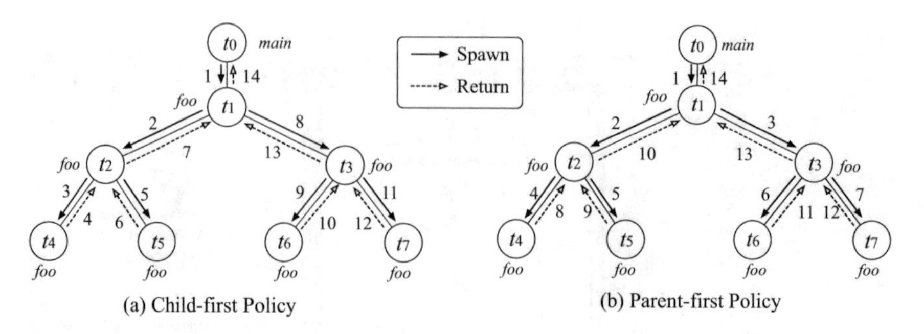

Fig. 3.10 Child-first policy verses Parent-first policy

return operations; and the number by side of each arrow is the order the operation is processed.

As shown in Fig. 3.10a, in the child-first policy, a worker executes the child task immediately after the child is spawned, leaving the parent task for later execution or for stealing by other cores. For instance, if child-first policy is adopted, the worker that executes t_1 immediately starts to execute t_2 when t_2 is spawned by t_1, and the worker immediately execute t_4 when it is spawned by t_2. In this way, if there is a single worker, the tasks are executed in the sequence of "$t_0 \rightarrow t_1 \rightarrow t_2 \rightarrow t_4 \rightarrow t_5 \rightarrow t_3 \rightarrow t_6 \rightarrow t_7$". Child-first policy is similar to the depth-first traversal of the task graph. For example, MIT Cilk uses the child-first policy, aka. work-first in [8]. Child-first policy works better when the steals are infrequent due to the light overhead [21].

As shown in Fig. 3.10b, in the parent-first policy, a worker continually executes the parent task after spawning a child task, leaving the child task for later execution or for stealing by other cores. For instance, if parent-first policy is adopted, the worker that executes t_1 continues to execute t_1 after spawned t_2. In this way, if there is a single worker, the tasks are executed in the sequence of "$t_0 \rightarrow t_1 \rightarrow t_2 \rightarrow t_3 \rightarrow t_4 \rightarrow t_5 \rightarrow t_6 \rightarrow t_7$". Parent-first policy is similar to the breadth-first traversal of the task graph. One example of the parent-first policy is the help-first policy described by Guo et al. [21]. Parent-first policy works better when the steals are frequent and the task graph is shallow [21], because the policy can quickly generates a large amount of tasks that can be distributed across cores.

Both the child-first and the parent-first task generating policies have their advantages and disadvantages. CAB uses both of the two policies in different situations to best utilize their advantages. In CAB, for an iterative program, during its first iteration, tasks have not been divided into inter-socket tasks and intra-socket tasks. For the convenience of collecting SOID, CAB adopts the parent-first policy in the first iteration. After the task graph has been divided into two tiers, CAB generates inter-socket tasks with the parent-first policy and generates intra-socket tasks with the child-first policy. CAB adopts the parent-first policy for generating inter-socket tasks so that leaf inter-socket tasks can be generated as soon as possible. The parent-first policy is more efficient in this case because inter-socket tasks take short time and thus are frequently stolen. On the other hand, CAB adopts the child-first policy to generate intra-socket tasks. The child-first policy works better in this case because the leaf tasks take longer time and thus the steals are infrequent. Also the child-first policy is more space efficient.

3.6 Theoretical Time and Space Bounds

In this section, we discuss the theoretical time bound and space bound of executing a parallel program, when it scheduled by CAB runtime system. In the following discussion, we model the execution of a parallel program as the traversal of a task graph G. Each node in G represents a unit task, and each edge represents a dependence between tasks.

3.6.1 Theoretical Bounds for Random Work-Stealing

The time and space bounds for random work-stealing have been proved by Blumofe et al. [9]. According to their discussion, for a task graph G, the *work* $T_1(G)$ is the number of nodes in G, and the critical-path length $T_\infty(G)$ is the number of nodes along the longest path from the start node to the end node.

For such task graph G, the time bound of executing G on a P-core computer, denoted by $T_P(G)$ can be calculated by Eq. 3.7. The work T_1 is the number of unit tasks in the task graph, and the critical path length T_∞ is the number of nodes along the longest path from the start node to the end node.

$$T_P(G) \leq \frac{T_1(G)}{P} + T_\infty(G) \tag{3.7}$$

For the same task graph G, the space used by G on a P-core computer, denoted by $S_P(G)$ can be calculated by Eq. 3.8. In this equation, $S_1(G)$ is the space used by the program when it is executed sequentially.

$$S_P(G) \leq S_1(G) \times P \tag{3.8}$$

If you are interested in understanding the details of the provement, please refer to the prior work [9]. The following discussion is based on the time and space bounds of the random work-stealing.

Table 3.2 Parameters used in the bound analysis

Parameters	Description
G_γ	The subtree rooted at task γ in G
m	Number of sockets
n	Number of cores per socket
c	Number of overall cores ($m \times n$)
$T_1(G)$	The total number of nodes in G
$T_\infty(G)$	The critical-path length of G
$T_P(G)$	Makespan of G on a P-core computer
$T_P(G_{inter})$	Makespan of the inter-socket tier
$T_P(G_{intra})$	Makespan of the intra-socket tier
$S_P(G)$	Space used by G on a P-core computer

3.6.2 Theoretical Bounds for CAB

Based on the time and space bounds of the traditional random work-stealing, we can analyze the time and space bounds of the cache-aware bi-tier work-stealing, CAB. Table 3.2 lists the parameters that are used to analyze G's time and space bounds.

3.6.2.1 Time Bound Analysis

Since A-CAB divides a DAG into two tiers and executes them differently, we need to divide a DAG into sub-DAGs using the leaf inter-socket tasks. Given a leaf inter-socket task t, we use the notation G_t to represent the subtree rooted with t, which includes the set of tasks that are generated from t. Therefore, the total work of G is divided as in Eq. 3.9, where G_{inter} represents the subgraph of the inter-socket tier and k is the total number of leaf inter-socket tasks.

$$T_1(G) = T_1(G_{inter}) + \sum_{i=1}^{k} T_1(G_{t_i}) \tag{3.9}$$

The execution time of G in an M-socket N-core architecture, $T_{M\times N}(G)$, can be divided into two parts: the execution time of the inter-socket tier $T_{M\times N}(G_{inter})$ and the execution time of the intra-socket tier $T_{M\times N}(G_{intra})$. Even though the two parts can be overlapped, we use their sum to get the worst bound of $T_{M\times N}(G)$ as shown in Eq. 3.10.

$$T_{M\times N}(G) = T_{M\times N}(G_{inter}) + T_{M\times N}(G_{intra}) \tag{3.10}$$

Since the inter-socket tier is executed by M head cores using work-stealing, according to the proof of [9], the execution time of G_{inter} is bounded by Eq. 3.11.

$$T_{M\times N}(G_{inter}) \leq \frac{T_1(G_{inter})}{M} + T_\infty(G_{inter}) \tag{3.11}$$

For the execution of the intra-socket tier, each G_{t_i} is executed by N cores within a socket using work-stealing. Therefore, the execution time of G_{t_i} is bounded by Eq. 3.12.

$$T_N(G_{t_i}) \leq \frac{T_1(G_{t_i})}{N} + T_\infty(G_{t_i}) \tag{3.12}$$

Since k leaf inter-socket tasks are scheduled among M sockets using work-stealing, the execution time of the intra-socket tier is bounded by Eq. 3.13.

$$T_{M\times N}(G_{intra}) \leq \frac{\sum_{i=1}^{k} T_N(G_{t_i})}{M} + T_\infty(G_{intra}) \tag{3.13}$$

Deducing from Eqs. 3.12 and 3.13, we can get Eq. 3.14.

$$T_{M\times N}(G_{intra}) \leq \frac{\sum_{i=1}^{k} T_1(G_{t_i})}{M\times N} + \frac{\sum_{i=1}^{k} T_\infty(G_{t_i})}{M} + T_\infty(G_{intra}) \tag{3.14}$$

From Eqs. 3.10, 3.11 and 3.14, $T_{M\times N}(G)$ can be bounded as in Eq. 3.15.

$$\begin{aligned} T_{M\times N}(G) \leq & \frac{T_1(G_{inter})}{M} + T_\infty(G_{inter}) + \frac{\sum_{i=1}^{k} T_1(G_{t_i})}{M\times N} + \\ & \frac{\sum_{i=1}^{k} T_\infty(G_{t_i})}{M} + T_\infty(G_{intra}) \end{aligned} \tag{3.15}$$

After further tidying Eq. 3.15 up, we have Eq. 3.16.

$$T_{M\times N}(G) \leq \frac{T_1(G_{inter})}{M} + \frac{T_1(G_{intra})}{M\times N} + \frac{\sum_{i=1}^{k} T_\infty(G_{t_i})}{M} + T_\infty(G) \tag{3.16}$$

Our experiments show that the execution time of the inter-socket tier is often less than 5% of the overall execution time. Therefore, the time bound of Eq. 3.16 is very close to the traditional random work-stealing schedulers such as MIT Cilk for many D&C applications.

3.6.2.2 Space Bound Analysis

According to the proof of [9], the space used by G in an M-socket N-core architecture is bounded by Eq. 3.17, where $S_1(G)$ denotes the space used by the serial execution of the program.

$$S_{M\times N}(G) \leq M\times N\times S_1(G) \tag{3.17}$$

Equation 3.17 assumes that there are at most $M \times N$ workers expanding the DAG using the child-first policy. However, since A-CAB uses the parent-first policy to expand the inter-socket tier quickly, each of the leaf inter-socket tasks may use S_1 space in the worst case. Therefore, the space used by the A-CAB scheduler $S_{M \times N}(G)$, can be bounded as in Eq. 3.18.

$$S_{M \times N}(G) \leq \max\{k \times S_1(G), M \times N \times S_1(G)\} \tag{3.18}$$

3.7 Implementation Methodology

CAB runtime system can be implemented in existing work-stealing programming environments. Without loss of generality, in the book, we use MIT Cilk that is one of the earliest work-stealing programming environments, as the baseline and implement CAB by modifying MIT Cilk. In this section, we first introduce the compiler support of CAB, including the support for FTO and GTO task graph partitioning techniques. After that, we discuss the runtime support for CAB. It is worthing noting that, Cilk programs can run in CAB runtime system without any modification.

3.7.1 *Compiler Support*

The compiler is modified to support both the parent-first and the child-first task-generating policy.

3.7.1.1 Support for FTO Task Graph Partition Technique

As we presented in Sect. 3.4.1, in the FTO task graph partition technique, a boundary level L_b is used to divide a task graph into two tiers. Therefore, when a new task is spawned, we first compare its level in the task graph L with the boundary level L_b. If $L \leq L_b$, then the task is in the inter-socket tier. In this case, the task is generated using parent-first policy. On the contrary, if $L > L_b$, the task is in the intra-socket tier and it is generated using child-first policy. Meanwhile, because inter-socket tasks and intra-socket tasks are generated in different policies, we modified the source-to-source compiler, Cilk2c, to support the two types of *sync*.

3.7.1.2 Support for GTO Task Graph Partition Technique

For a non-iterative program, at each spawn, if the to-be-spawned task's SOID is smaller than the size of the shared cache, CAB spawns the task with the child-first policy and pushes the task into the intra-socket task pool of the current core.

Otherwise, CAB spawns the task with the parent-first policy and pushes the task into the inter-socket task pool of the current socket. For an iterative program, at each spawn, CAB finds out whether the spawn happens in the first iteration of the program. If it is in the first iteration, the to-be-spawned task is spawned with the parent-first policy. Otherwise, the to-be-spawned task is spawned and scheduled in the same way as the tasks in non-iterative program.

Since the compiling-based method needs the branching degrees of all the tasks, we have also modified *cilk2c* to acquire them by analyzing the source code based on the keyword *spawn*. Additionally, for non-iterative programs, we have further modified *cilk2c* to insert instructions that compute and record the SOID of each and every task according to Eq. 3.5 when the task is spawned.

3.7.2 Runtime Support

If a parallel program runs on an M-socket architecture, in which each socket has N-cores, CAB runtime system launches $M \times N$ workers and affiliates each worker with an individual hardware core. For each socket, the worker that affiliated with core "0" is the header worker of the socket. The worker is responsible to balance the workload among different sockets.

Table 3.3 gives the algorithm used to schedule tasks in a parallel program in CAB runtime system. Note that, before CAB starts to schedule the tasks in a parallel program, we already know the task graph is to be divided into two tiers using FTO policy or GTO policy, and know whether the program is iterative or non-iterative.

3.8 Evaluation of CAB

In this section, we use a Dell 16-core computer that has four AMD Quad-core Opteron 8380 processors (codenamed "Shanghai") running at 2.5 GHz to evaluate the performance of CAB runtime system. Each Quad-core socket has a 512 K private L2 cache for each core and a 6M L3 cache shared by all four cores. The computer has 16GB RAM and runs Linux 2.6.29. Accordingly, CAB sets up four workers in each socket.

In this chapter, we described two policies: FTO policy and GTO policy to divide a task graph into two tiers. In this section, we evaluate the performance of the two policies in CAB runtime system. For easy of description, we use CAB-FTO to represent the CAB runtime system that employs FTO task graph partition policy; and use CAB-GTO to represent the CAB runtime system that employs GTO task graph partition policy. In Sects. 3.8.1 and 3.8.2, we present evaluation results for CAB-FTO and CAB-GTO respectively.

As described earlier, in CAB runtime system, each worker thread is affiliated with a hardware core. Our experiment shows that affiliating workers with cores can improve the performance of memory-bound applications in CAB (shown in Figs. 3.11

Table 3.3 CAB runtime algorithm

Assumption: Suppose an M-socket and N-core architecture and a worker w belongs to a socket ρ.
Global initiation:
Step 1: CAB launches $M \times N$ workers and affiliates them to the corresponding hardware cores
Step 2: CAB identifies which policy (either FTO policy or GTO policy) is to be used divided the task graph into inter-socket tier and intra-socket tier
Step 3: If FTO policy is used, CAB calculates the boundary level L_b. If M equals 1, CAB sets L_b to 0. Otherwise, CAB calculates L_b according to Eq. 3.4
Step 4: Worker 0 begins to execute the initial task, while all the other workers are trying to steal tasks
Task scheduling: Assume worker w in socket ρ is executing task t.
(a) t **generates** t_1:
①. If FTO policy is adopted, task t first calculates which level task t_1 is in. If t_1 is in inter-socket tier, it is an inter-socket task. In this case, w pushes t_1 into the inter-socket task pool of socket ρ, and continues to execute task t. On the other hand, if t_1 is in intra-socket tier, it is an intra-socket task. In this case, w pushes t_1 into the intra-socket task pool of w, and starts to execute task t_1 immediately
②. If GTO policy is adopted and the program is non-iterative, t calculates the SOID of task t_1 according to Eq. 3.6, identifies which level t_1 is in, and executes tasks as described in ①
③. If GTO policy is adopted and CAB is executing the first iteration of an iterative program, then w pushes the task t_1 into the inter-socket task pool of ρ, and continues to execute task t
④. If GTO policy is adopted and CAB is executing the following iterations of a program, t searches the SOID of t_1, identifies which tier t_1 is in, and executes tasks as described in ①
(b) t **suspends**: w tries to obtain a task according to the algorithm described in Fig. 3.9
(c) t **returns**: w returns the results of t and sets *busy_state* of ρ to false if t is an inter-socket task. Then w tries to get a task according to the algorithm described in Fig. 3.9
Termination: If all the tasks have finished, sCAB terminates

and 3.17). In the default MIT Cilk, the workers are not affiliated with hardware cores. For the fairness of comparison, Therefore, we have modified the MIT Cilk (denoted as *Cilk* for short) to affiliate each worker with a hardware core (denoted as *Cilk-a* for short).

In all of our experiments, Cilk-a uses the pure child-first policy to spawn and schedule tasks, while CAB-FTO and CAB-GTO flexibly use both the child-first and parent-first policies to achieve the best performance. We implement the evaluated work-stealing runtime schedulers based on MIT Cilk. The MIT Cilk programs run with Cilk-a, CAB-FTO, and CAB-GTO without any modification. All benchmarks are compiled with "cilk2c -O2", which is based on gcc 4.4.3. Furthermore, for each test, every benchmark is run ten times, and the average execution time is used in the final results.

3.8.1 *Performance of CAB-FTO*

Table 3.4 lists the benchmarks used to evaluate the performance of CAB-FTO. Because CAB-FTO only works for parallel programs that have full tree task graphs, the task graphs of all the memory-bound benchmarks in Table 3.4 are full tree-shaped. In Sect. 3.8.2, benchmarks that have irregular task graphs are used to evaluate the performance of CAB-GTO that works for general applications.

3.8.1.1 Performance of CAB-FTO for Memory-Bound Programs

Figure 3.11 presents the normalized execution time of memory-bound benchmarks with a 1024 $\times$ 1024 matrix as input dataset in Cilk-a, Cilk, and CAB-FTO. Observed from the figure, CAB-FTO significantly improves the performance of memory-bound benchmarks compared with Cilk and Cilk-a, while the performance improvements ranges from 10 to 55%. As shown in Fig. 3.11, because Cilk-a always performs better than Cilk in our experiment, only the performance of Cilk-a and CAB-FTO is compared in the following chapter.

The main reason that CAB-FTO can improve the performance of memory-bound programs is that CAB-FTO relieves the TRCI problem described in Sect. 3.1.1. Table 3.5 lists the L2 cache (private) misses and L3 cache (shared) misses of the memory-bound programs in Cilk-a, and CAB-FTO. From the table, we can find that CAB-FTO can significantly reduce both the L2 cache misses and L3 cache misses. This is mainly because CAB-FTO schedules the tasks that share data into the same socket, and the dataset of the tasks running in the same socket can fit in the L3 cache of a socket. On the contrary, because Cilk-a randomly schedules tasks to the workers, it increases the memory footprint and thus increase the L3 shared cache misses.

Table 3.4 Benchmarks used to evaluate CAB-FTO

Name	Type	Description
Queens(20)	CPU-bound	N-queens problem
FFT	CPU-bound	Fast fourier transform
CK	CPU-bound	Rudimentary checkers
Cholesky	CPU-bound	Cholesky decomposition algorithm
Heat	Memory-bound	Five-point heat distribution algorithm
SOR	Memory-bound	Successive over-relaxation
GE	Memory-bound	Gaussian elimination algorithm
Mergesort	Memory-bound	Mergesort on 1024 $\times$ 1024 Integers

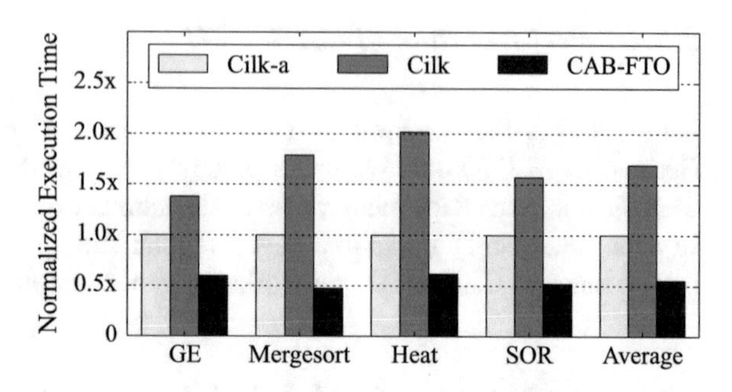

Fig. 3.11 The normalized execution time of memory-bound benchmarks with a 1024 × 1024 matrix as the input dataset in Cilk-a, Cilk, and CAB-FTO

Table 3.5 L2 cache misses and L3 cache misses of memory-bound benchmarks in Cilk-a and CAB-FTO

	Scheduler	GE	Mergesort	Heat	SOR
L2 cache misses	Cilk-a	4203604	5717785	8457899	14134418
	CAB-FTO	2617207	3448768	5577723	10863876
L3 cache misses	Cilk-a	1545310	1974802	2812464	5259771
	CAB-FTO	180145	998605	755786	1256203

Because CAB-FTO reduces the cache misses of benchmarks compared with Cilk-a, it performs much better than Cilk-a.

3.8.1.2 Effectiveness of the FTO

In the FTO task graph partition policy (Sect. 3.4.1), the boundary level L_b, which divides a task graph into two tiers, is calculated using user-provided parameters. In this subsection, we use benchmark *Heat* to evaluate the effectiveness of calculating the boundary level L_b using Eq. 3.4. Since other benchmarks show similar results, we only present the experimental results of *Heat* here.

In this experiment, we evaluate the performance of *Heat* with all possible L_b values. Since *Heat* divides the input dataset into two parts each time sub-tasks are generated until the data size smaller than 64 rows, there are fewer possible L_b values when the input datasets are small.

Figure 3.12 shows the execution time of *Heat* with different input dataset size and all possible L_b value. For instance, when the input dataset is a $3k \times 2k$ matrix of *double*, there are 8 levels (0–7) in the task graph of *Heat*. In addition, because each item in the matrix is 8 bytes and *Heat* uses two copies of the input matrix, the overall input dataset size is $3072 \times 2048 \times 8 \times 2 = 96MB$. According to Eq. 3.4, the boundary level $L_b = \max\{\lceil \log_2 4 + 1 \rceil, \lceil \log_2 (96MB/6MB) + 1 \rceil\} = 5$. Observed from Fig. 3.12, when the input dataset is a $3k \times 2k$ matrix, *Heat* achieves the best performance when $L_b = 5$. The L_b values calculated for other data sizes are the ones

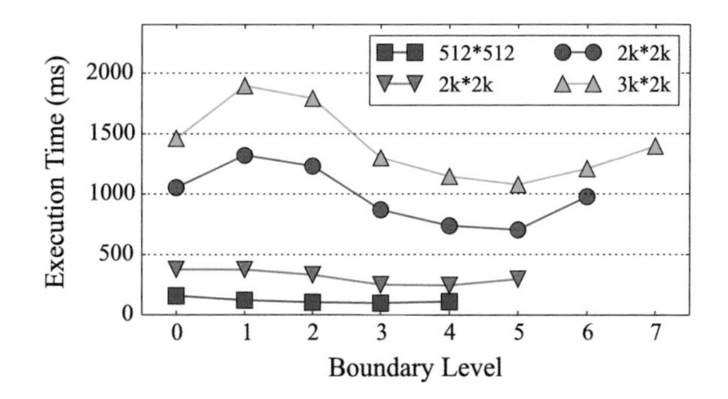

Fig. 3.12 Impact of L_b on the performance of *Heat* with different input dataset size. The FTO task graph partition method can find the best value for L_b

with the best performance as well according to Fig. 3.12. This proves the effectiveness of Eq. 3.12 and the FTO task graph partition method.

Note that, for larger data sizes, when L_b is smaller than three, the performance of CAB-FTO is poor. This is because, when L_b is small, there are only a small number of leaf inter-socket tasks. In this situation, workload is not balanced well in CAB-FTO, because it may not utilize all the sockets due to the lack of inter-socket tasks. One such extreme case is when $L_b = 1$, there is only one leaf inter-socket task, and thus only workers in one socket can get the task.

On the other hand, if L_b is too large (e.g., > 5), each leaf inter-socket task only contains a small number of intra-socket tasks. In this situation, the workload within a socket cannot be balanced well in CAB-FTO. For instance, for $L_b = 7$ in the case of $3k \times 2k$, leaf inter-socket tasks are in level 7 and do not generate any intra-socket tasks. In this case, there is only one worker contributing to the performance of every socket.

3.8.1.3 Scalability of CAB-FTO

The size of input dataset affects the performance of memory-bound programs in CAB-FTO. If the input dataset is large, the performance improvement in CAB-FTO is relatively small. In this experiment, we use benchmarks *Heat* and *SOR* as examples to evaluate the scalability of CAB-FTO. Experiments on other memory-bound benchmarks show similar results.

Figure 3.13 shows the performance of *Heat* and *SOR* with different input dataset size. Observed from this figure, for *Heat*, when the input dataset is small (512×512), the performance of *Heat* is improved by 54.6%; but when the input dataset is large ($4k \times 4k$), the improvement drops to 14%. Meanwhile, for *SOR*, when the input dataset is small (512×512), the performance of *SOR* is improved by 68.7%; but when the input dataset is large ($4k \times 4k$), the improvement drops to 13.6%.

One reason for the diminishing improvement is that, with the increasing input data sizes, the shared data set between intra-socket tasks becomes relatively smaller, which increases the proportion of non-shared data and the cache misses in the leaf inter-socket tasks. Figure 3.14 shows the shared cache misses of *Heat* and *SOR* with different input dataset size. Observed from the figure, CAB-FTO can reduce 68.2%

Fig. 3.13 Performance of heat and SOR with different input dataset size

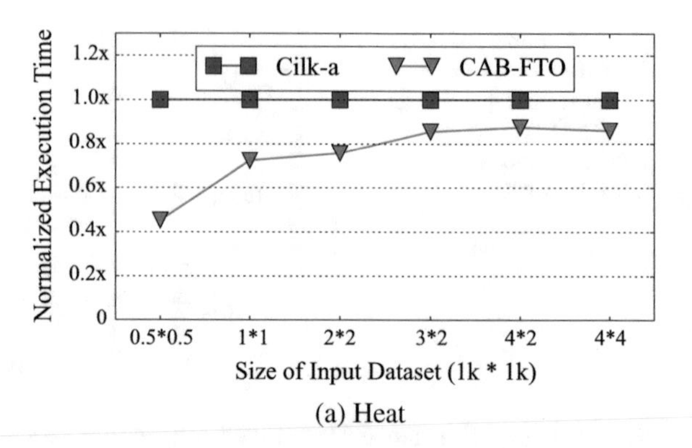

(a) Heat

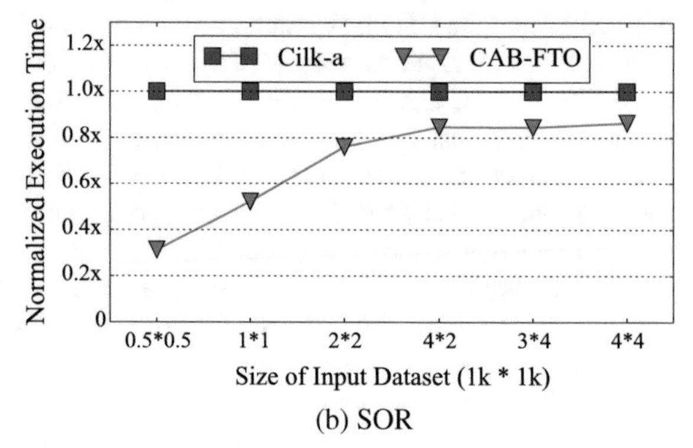

(b) SOR

shared cache misses when the input dataset is small, and reduce 10.6% shared cache misses when the input dataset is large, compared with Cilk-a.

Another reason for the diminishing improvement is that, when the input dataset is large, the granularity of the leaf tasks becomes large, which results in fewer intra-socket tasks in an intra-socket subtree and it is not good for load balance within a socket.

3.8.1.4 Performance of CAB-FTO for CPU-Bound Programs

Because the main purpose of CAB-FTO is to relieve the TRICI problem for memory-bound programs, CPU-bound programs cannot achieve better performance in CAB-FTO compared with Cilk-a. For a CPU-bound program, CAB-FTO can use traditional random work-stealing policy to schedule the tasks. In the implementation of CAB-FTO, by setting the boundary level $L_b = 0$, random work-stealing policy is adopted to schedule tasks.

Figure 3.15 shows the performance of CPU-bound benchmarks in Table 3.4 in Cilk-a and CAB-FTO. Observed from the figure, in most cases, the extra-overhead caused by CAB-FTO is between 1 to 2% of the overall execution time, which is often negligible.

Fig. 3.14 L3 shared cache misses of heat and SOR in CAB-FTO and Cilk-a

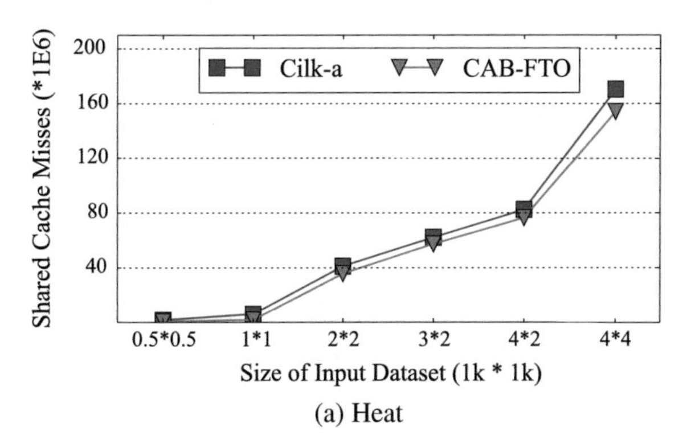

(a) Heat

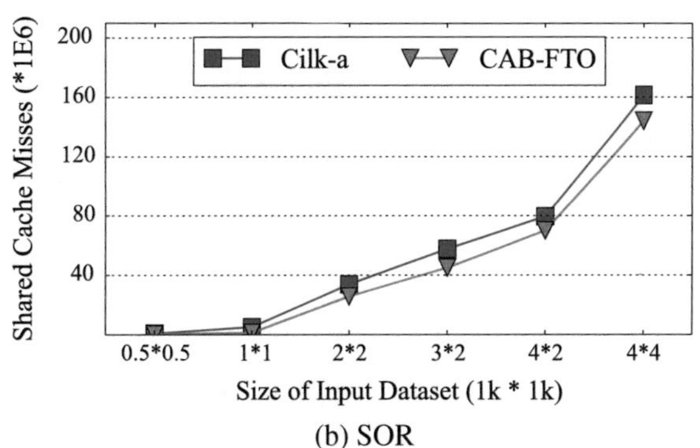

(b) SOR

Fig. 3.15 The normalized execution time of CPU-bound benchmarks in Cilk-a, and CAB-FTO

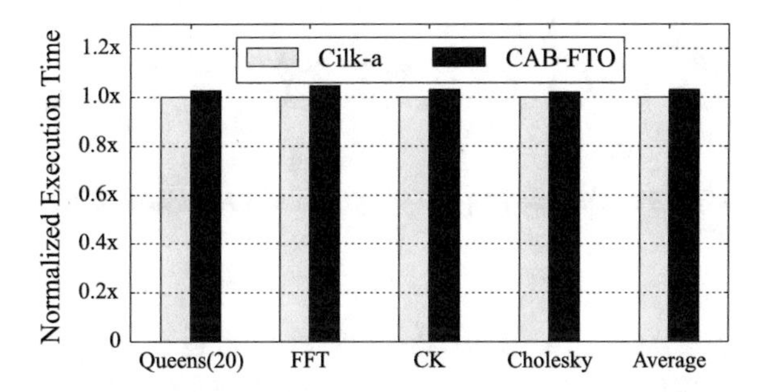

3.8.2 Performance of CAB-GTO

Similar to Sect. 3.8.1, we also use memory-bound benchmarks to evaluate the performance of CAB-GTO. CPU-bound benchmarks are used to measure the extra overhead of CAB-GTO compared with random work-stealing.

In order to evaluate the performance of CAB-GTO in different scenarios, we use benchmarks that have both regular (full tree-shaped) and irregular task graphs in the experiments. Table 3.6 lists the benchmarks used to evaluate the performance of CAB-GTO. Since we can configure the iteration number of the memory-bound

Table 3.6 Benchmarks used to evaluate CAB-GTO

Name	Type	Description
Mandelbrot (MB)	CPU-bound	Calculate mandelbrot set
Queens (15)	CPU-bound	N-queens problem
FFT	CPU-bound	Fast fourier transform
GA	CPU-bound	Island Model of Genetic Algorithm
Knapsack	CPU-bound	0–1 knapsack problem
Heat	Memory-bound	Five-point heat
Heat-ir	Memory-bound	Five-point heat (irregular)
SOR	Memory-bound	Successive over-relaxation
SOR-ir	Memory-bound	Successive over-relaxation (irregular)
GE	Memory-bound	Gaussian elimination
GE-ir	Memory-bound	Gaussian elimination (irregular)

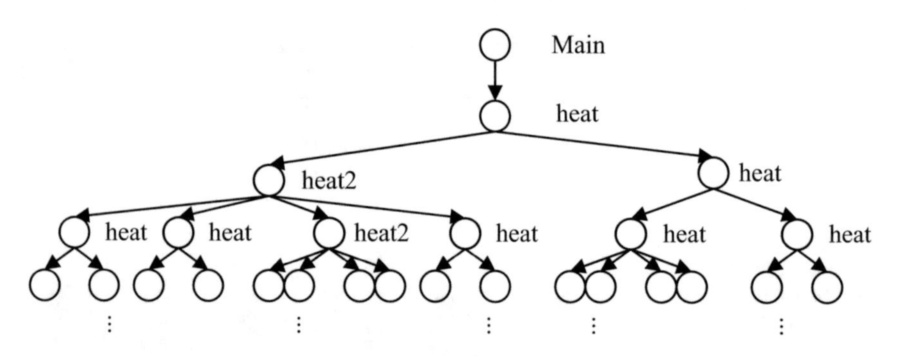

Fig. 3.16 The task graph of the program in Algorithm 3. This task graph is an irregular non-full tree

benchmarks, we can evaluate the compiling-based method for non-iterative programs by adjusting the benchmarks to run only one iteration.

Heat-ir, *GE-ir* and *SOR-ir* implement the same algorithm as *Heat*, *GE* and *SOR* respectively, except their task graphs are unbalanced trees. For example, we implement *Heat-ir* in Algorithm 3. According to the algorithm, the branching degree of tasks created from cilk procedure *heat* is 2 while the branching degree of tasks created from cilk procedure *heat2* is 4. Figure 3.16 shows the task graph of the program in Algorithm 3. Obviously, the task graph is irregular. *GE-ir* and *SOR-ir* are implemented in the similar way.

Algorithm 3 The source code skeleton of *Heat-ir*.

```
cilk void heat (int start, int end) {
  if(end-start < threshold) {
    Process data ;
  } else {
    int mid = (start + end) / 2;
    spawn heat2 (start, mid);
    spawn heat (mid, end);
    sync;
    return;
  }
}
cilk void heat2 (int start, int end) {
  if(end-start < threshold) {
    Process data ;
  } else {
    int quad = (end - start) / 4;
    spawn heat (start, start + quad);
    spawn heat (start + quad, start + 2*quad);
    spawn heat2 (start + 2*quad, start + 3*quad);
    spawn heat (start + 3*quad, end);
    sync;
    return;
  }
}
cilk int main (int start, int end) {
  ...
  spawn heat (start, end);
  sync;
  ...
  return 0;
}
```

3.8.2.1 Performance of CAB-GTO for Memory-Bound Benchmarks

Iterative memory-bound benchmarks. We first evaluate the performance of CAB-GTO for iterative memory-bound parallel programs. Figure 4.10 shows the performance of memory-bound benchmarks in Cilk, Cilk-a and CAB-GTO with a 1024×512 matrix as the input data. For *GE* and *GE-ir*, the used input data is a 1024×1024 matrix due to the algorithm requirement. All the benchmarks consist of 200 iterations in this experiment. Since the benchmarks are iterative, CAB-GTO adopts the profiling-based method to calculate the SOIDs of tasks at the first iteration of the benchmarks.

As we can see from Fig. 3.17, CAB-GTO with the profiling-based method can significantly improve the performance of iterative memory-bound benchmarks compared to Cilk-a while the performance improvement ranges from 35.3 to 74.4%. On average, the performance improvement is up to 43.7% compared with Cilk-a.

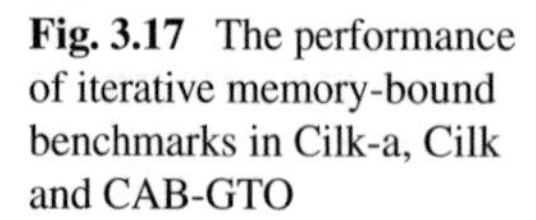

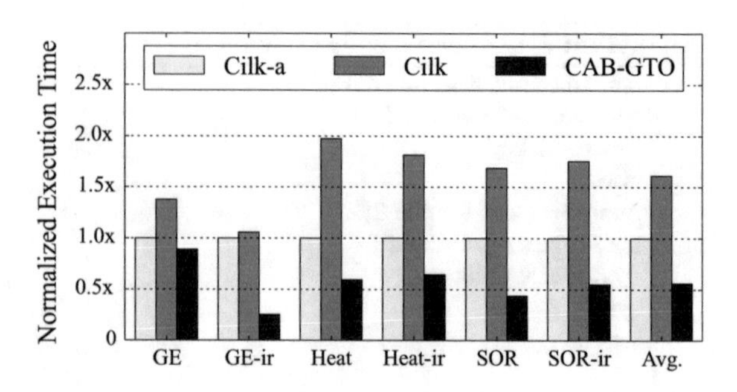

Fig. 3.17 The performance of iterative memory-bound benchmarks in Cilk-a, Cilk and CAB-GTO

Table 3.7 Cache misses in Cilk-a and CAB-GTO ($\times$1E6)

Benchmark	Scheduling system	L1 cache miss	L2 cache miss	L3 cache miss
GE	Cilk-a	60.8	58.8	14.5
	CAB-GTO	53.9	50.3	2.94
GE-ir	Cilk-a	37.2	37.1	10.7
	CAB-GTO	23.9	20	2.15
Heat	Cilk-a	82.7	79.6	24.8
	CAB-GTO	71.1	67.5	5.9
Heat-ir	Cilk-a	82.2	78.7	29.7
	CAB-GTO	71.3	67.6	3.72
SOR	Cilk-a	88.5	85	29.6
	CAB-GTO	70.7	66.2	4.75
SOR-ir	Cilk-a	89.8	85.5	30.7
	CAB-GTO	73.6	67.4	8.27

Similar to Sect. 3.8.1, in order to explain why CAB-GTO can improve the performance of iterative memory-bound programs compared with Cilk-a, we collect the cache misses of all the benchmarks and list them in Table 3.7. Observed from the table, we can find that the shared cache (L3) misses are prominently reduced while the private cache (L1 and L2) misses are also slightly reduced in CAB-GTO compared with Cilk-a. Since CAB-GTO schedules tasks with shared data into the same socket, the shared cache misses have been significantly reduced.

Although scheduling tasks with shared data to the same socket only reduces the shared L3 cache misses, the affiliation of an intra-socket subtree with a socket in CAB-GTO can help reduce the private L1 and L2 cache misses slightly. In CAB-GTO, for a task t_i in an intra-socket subtree, if it is executed by core c in socket ρ, its neighbor tasks (i.e., t_{i-1} and t_{i+1}) are also executed by c as well unless they are stolen by other cores in ρ. Compared with random work-stealing where any free cores can steal t_i's neighbor tasks, there are fewer cores that can steal t_i's neighbor tasks in CAB-GTO. Therefore, the probability that neighbor tasks are executed by

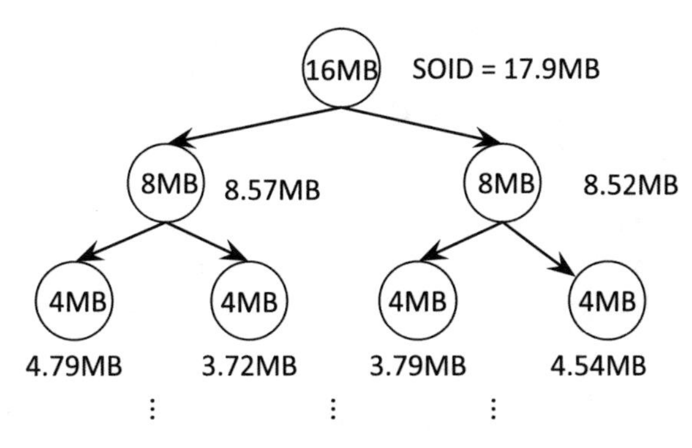

Fig. 3.18 SOIDs of tasks in *Heat* with a 1024×512 matrix as input data

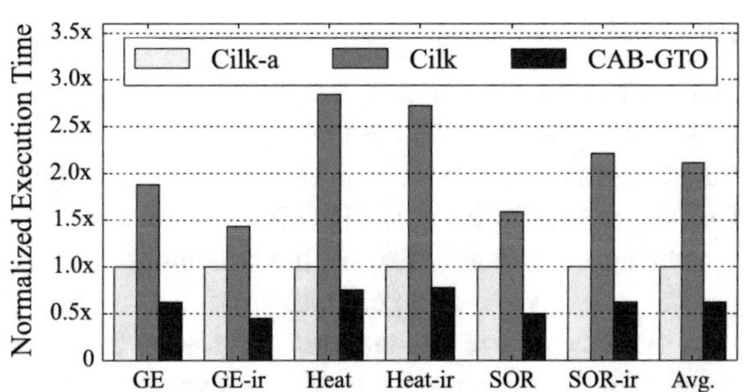

Fig. 3.19 The performance of non-iterative memory-bound benchmarks in Cilk-a, Cilk and CAB-GTO

the same core is larger in CAB-GTO. For this reason, the private cache (e.g., L1 and L2) misses have also been slightly reduced in CAB-GTO.

Take benchmark *Heat* with a 2048×512 matrix as input data as an example. Figure 3.18 shows the SOIDs of tasks in *Heat* that are calculated with Eq. (3.6). The real involved data size of tasks in Fig. 3.18 are shown in the circles. Since *Heat* uses two matrices of "*double*" during the execution, the overall input data size is $2048 \times 512 \times 8 \times 2 = 16$ MB. Then the real data set is evenly divided every time when the tasks are spawned. From the figure, we can find that the calculated SOIDs are close to the real involved data sizes, which shows the profiling-based method is reasonably accurate.

Non-iterative memory-bound benchmarks. We evaluate the performance of CAB-GTO for non-iterative memory-bound parallel programs. Figure 3.19 shows the performance of non-iterative benchmarks in Cilk, Cilk-a and CAB-GTO. We create the non-iterative benchmarks by setting their iteration number as 1. CAB-GTO adopts the compiling-based method to calculate the SOIDs of tasks for non-iterative programs.

With the compiling-based method, CAB-GTO significantly improves the performance of non-iterative memory-bound applications compared to Cilk-a while the performance improvement ranges from 22.3 to 55.1%. On average, the performance improvement is up to 38.1% compared with Cilk-a. Same as iterative benchmarks, the good performance of non-iterative applications in CAB-GTO origins from the reduced shared cache misses as well.

From Figs. 3.17 and 3.19 we can find that Cilk-a provides much better performance compared with Cilk for all the benchmarks. For memory-bound applications, the better performance in Cilk-a results from the affiliation of the workers with the cores. In the rest of our experiments, we only compare the performance of CAB-GTO with Cilk-a.

In summary, with the profiling-based method for iterative programs and the compiling-based method for non-iterative programs, CAB-GTO is effective for memory-bound divide-and-conquer programs.

3.8.2.2 Scalability of CAB-GTO

In order to evaluate the scalability of CAB-GTO in different scenarios, benchmarks that have both regular and irregular task graphs are used in this section. In this experiment, we can change the input data sizes of the benchmarks. By comparing the performance of benchmarks with different input data sizes in CAB-GTO and Cilk-a, the scalability of CAB-GTO is evaluated.

During the execution of all the benchmarks, every task divides its data set into several parts by rows to generate child tasks unless the task meets the cutoff point (i.e., the data set size of a leaf task). Since the data set size of the leaf tasks affects the measurement of scalability, the data set size of the leaf tasks in the experiment should be constant. To satisfy this requirement, a constant cutoff point, 8 rows, is used for the leaf tasks, and a constant number of columns, 512, for the input data. Only the number of rows of the input matrix is adjusted in the experiment. In this way, the scalability of CAB-GTO can be measured without the impact of the granularity of the leaf tasks. In all the following figures, the x-axis represents the number of rows of the input matrix.

Benchmarks with Regular Task Graphs *Heat* and *SOR* are used as benchmarks to evaluate the scalability of CAB-GTO for applications with regular task graphs. Other benchmarks, such as *GE*, have similar results.

Figure 3.20 shows the performance of *Heat* and *SOR* with different input data sizes in Cilk-a and CAB-GTO. From Fig. 3.20, we can see that *Heat* and *SOR* achieve better performance in CAB-GTO for all sizes of the input data up to 8192 rows compared with Cilk-a. When the input data size is small (i.e., 1024×512), CAB-GTO reduces 40.4% execution time of *Heat* and reduces 56.1% execution time of *SOR*. When the input data size is large (i.e., 8192×512), CAB-GTO reduces 12.3% execution time of *Heat* and reduces 21.1% execution time of *SOR*.

Figure 3.21 shows the L2 and L3 cache misses of *Heat* with different input data sizes in Cilk-a and CAB-GTO. Observed from the figure, we can find that both the shared cache misses and the private cache misses are reduced in CAB-GTO compared with Cilk-a. The better performance of *Heat* in CAB-GTO results from the less cache misses in CAB-GTO compared with Cilk-a. When the input data size is small (1024×512), CAB-GTO can reduce 76.1% L3 cache misses and 15.2% L2 cache misses compared with Cilk-a. When the input data size is large (8192×512), CAB-GTO can reduce 55.9% L3 cache misses and 3.6% L2 cache misses compared

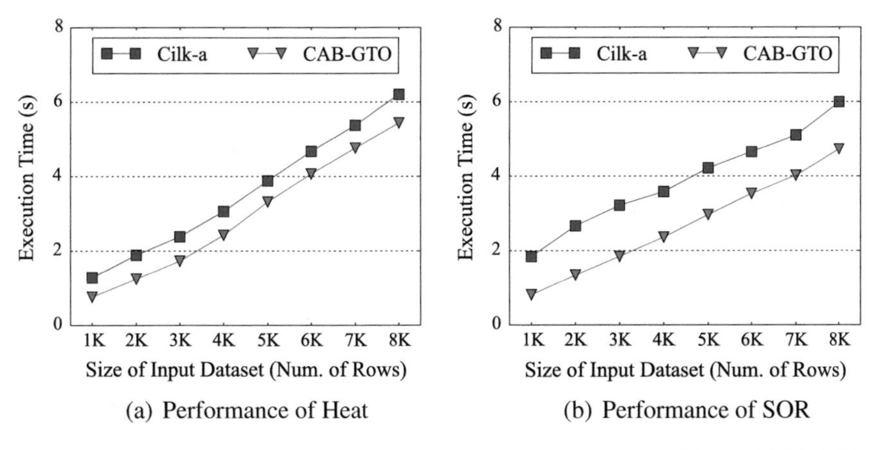

Fig. 3.20 Performance of Heat and SOR with different input data sizes in Cilk-a and CAB-GTO

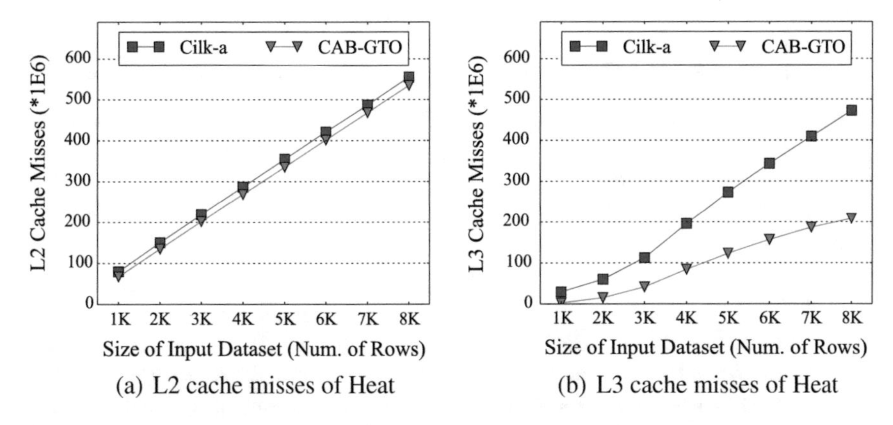

Fig. 3.21 L2 and L3 cache misses of Heat with different input data sizes in Cilk-a and CAB-GTO

with Cilk-a. Therefore, when CAB-GTO schedules regular memory-bound programs with regular task graphs, it is scalable. Other benchmarks show similar results of cache misses.

Benchmarks with Irregular Task Graphs *Heat-ir* and *SOR-ir* are used as benchmarks to evaluate the scalability of CAB-GTO for memory-bound programs with irregular task graphs. Other benchmarks, such as *GE-ir*, have similar results.

Figure 3.22 shows the performance of *Heat-ir* and *SOR-ir* with different input data sizes in Cilk-a and CAB-GTO. From Fig. 3.22 we can find that *Heat-ir* and *SOR-ir* also achieve better performance in CAB-GTO for all input data sizes compared with Cilk-a. When the input data size is small (i.e., 1024×512), CAB-GTO reduces 35.3% execution time of *Heat-ir* and reduces 44.9% execution time of *SOR-ir*. When the input data size is large (i.e., 8192×512), CAB-GTO reduces 11.4% execution time of *Heat-ir* and reduces 18% execution time of *SOR-ir*.

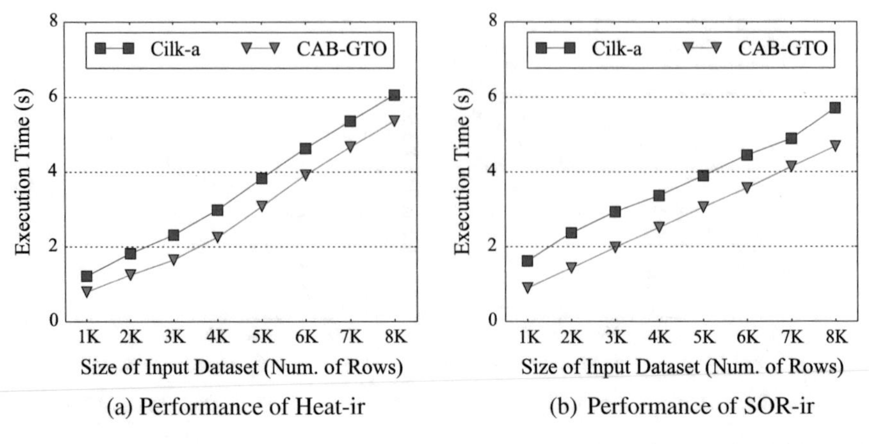

Fig. 3.22 Performance of Heat-ir and SOR-ir with different input data sizes in Cilk-a and CAB-GTO

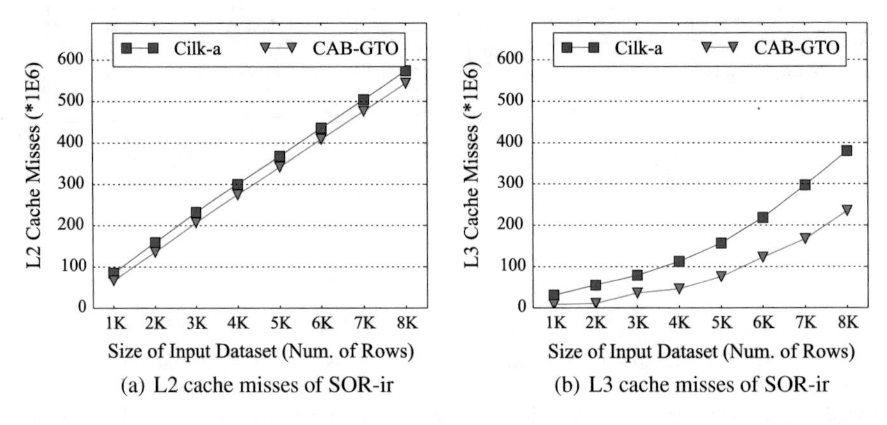

Fig. 3.23 L2 and L3 cache misses of SOR-ir with different input data sizes in Cilk-a and CAB-GTO

Figure 3.23 shows the L2 and L3 cache misses of *SOR-ir* with different input data sizes. Observed from the figure, we can find that both the shared cache misses and the private cache misses of *SOR-ir* are reduced in CAB-GTO compared with Cilk-a. The better performance of *SOR-ir* in CAB-GTO results from the less cache misses in CAB-GTO compared with Cilk-a. When the input data size is small, CAB-GTO can reduce 73.1% L3 cache misses and 21.2% L2 cache misses compared with Cilk-a. When the input data size is large, CAB-GTO can reduce 38.2% L3 cache misses and 5.2% L2 cache misses compared with Cilk-a. Other benchmarks show similar results of cache misses.

As illustrated in Figs. 3.20 and 3.22, the execution times of the benchmarks in both Cilk-a and CAB-GTO increase linearly to the input data size, because the execution times of the memory-bound benchmarks in both Cilk-a and CAB-GTO are determined by the input data size. However, for all the input data sizes, CAB-

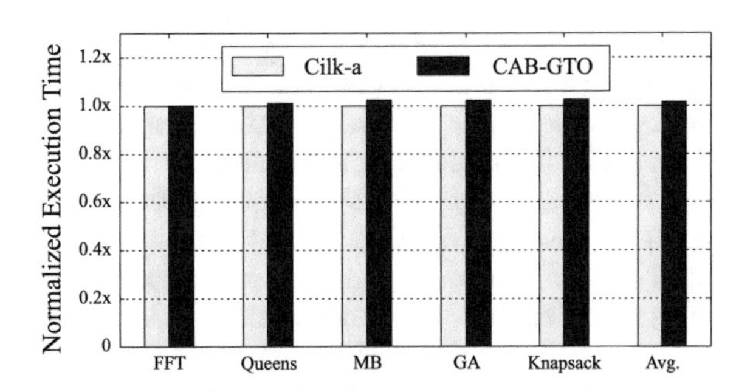

Fig. 3.24 Performance of CPU-bound benchmarks in Cilk-a and CAB-GTO

GTO can reduce the execution times of the memory-bound benchmarks accordingly. Therefore, CAB-GTO is scalable in scheduling both regular task graphs and irregular task graphs.

In addition, Figs. 3.21 and 3.23 further verify that CAB-GTO can also slightly reduce private cache misses by scheduling tasks with shared data into the same socket, which is due to the same reason explained previously.

In summary, the experiment in this section shows that CAB-GTO is scalable. It improves the performance of memory-bound benchmarks by reducing shared L3 cache misses. The experiment also shows that CAB-GTO can slightly reduce the private L2 cache misses, which is the by-product of the bi-tier work-stealing algorithm in CAB-GTO.

3.8.2.3 Performance of CAB-GTO for CPU-Bound Benchmarks

Since CAB-GTO is proposed to reduce shared cache misses of memory-bound applications, it is neutral to CPU-bound applications. Therefore, for CPU-bound applications, CAB-GTO uses traditional random work-stealing to schedule the tasks as Cilk-a.

Figure 4.16 shows the performance of CPU-bound benchmarks listed in Table 3.6 in Cilk-a and CAB-GTO. By comparing the performance of CAB-GTO with Cilk-a, we can find the extra overhead of CAB-GTO. Observed from Fig. 4.16, we see the extra overhead of CAB-GTO is negligible compared with Cilk-a. The extra overhead of CAB-GTO mainly comes from the profiling overhead when CAB-GTO can determine if the program is CPU-bound or memory-bound based on the profiling information.

3.9 Summary

In this chapter, we discuss state-of-the-art work-stealing scheduling policy for emerging multi-socket architecture. In multi-socket architecture, If the two tasks with shared data are scheduled to the same socket, only one of them needs to read the shared data from the main memory while the other task can access the data from the shared cache directly. Based on this observation, a Cache-Aware Bi-tier (CAB) work-stealing policy that automatically schedules tasks with shared data into the same socket is introduced in this chapter.

CAB is targeting divide-and-conquer (D&C) memory-bound applications, which covers a wide range of scientific applications in fluid dynamics, quantum dynamics, binary alloys, electromagnetism, superconductivity, thermodynamics, environmental systems, etc. The simulation of these systems has data parallelism that is often exploited with stencil-based approaches [4]. These applications are often iterative since the computation is repeated in steps of time. Their task graphs in systems like MIT Cilk are tree-shaped, which is ideally suitable for CAB.

CAB consists of a **cache aware task graph partitioner** and a **bi-tier work-stealing scheduler**. The task graph partitioner divides the task graph of a parallel program into the inter-socket tier and the intra-socket tier. The bi-tier work-stealing scheduler allows tasks in the inter-socket tier to be stolen across sockets, while tasks in the intra-socket tier are scheduled within the same socket. Since tasks from the intra-socket tier often share data, CAB uses the shared cache efficiently. Experimental results demonstrate that CAB significantly reduces the shared cache misses and thus improves the performance of memory-bound applications. The experiment shows that CAB can achieve a performance gain of up to 74.4% compared with traditional work-stealing.

3.9.1 Chapter Highlights

The following highlights of this chapter could be of your interest:

- We systematically analyze the TRICI problem caused by the random work-stealing in MSMC architecture.
- We propose a *profiling-based method* and a *compiling-based method* that collect the data access feature of tasks for iterative programs and non-iterative programs respectively. Based on the collected information, a DAG partitioner optimally divides tasks into the inter-socket tier and the intra-socket tier.
- We propose a bi-tier work-stealing algorithm that schedules tasks with shared data to the same socket.
- We demonstrate that CAB significantly reduces the shared cache misses and thus improves the performance of memory-bound applications. The experiment shows that CAB can achieve a performance gain of up to 74.4% compared with traditional work-stealing.

References

1. U. Acar, G. Blelloch, and R. Blumofe. The data locality of work stealing. Theory of Computing Systems, 35(3):321–347, 2002.
2. E. Ayguadé, N. Copty, A. Duran, J. Hoeflinger, Y. Lin, F. Massaioli, X. Teruel, P. Unnikrishnan, and G. Zhang. The design of openmp tasks. IEEE Transactions on Parallel and Distributed Systems, 20(3):404–418, 2009.
3. R. Azimi, M. Stumm, and R. Wisniewski. Online performance analysis by statistical sampling of microprocessor performance counters. In *Proceedings of the 19th annual international conference on Supercomputing*, pages 101–110. ACM, 2005.
4. M. Berger and J. Oliger. Adaptive mesh refinement for hyperbolic partial differential equations. Journal of computational Physics, 53(3):484–512, 1984.
5. G. Blelloch, R. Chowdhury, P. Gibbons, V. Ramachandran, S. Chen, and M. Kozuch. Provably good multicore cache performance for divide-and-conquer algorithms. In *Proceedings of the 19th annual ACM-SIAM symposium on Discrete algorithms*, pages 501–510. Society for Industrial and Applied Mathematics, 2008.
6. G. Blelloch, J. Fineman, P. Gibbons, and H. V. Simhadri. Scheduling irregular parallel computations on hierarchical caches. In *Proceedings of the 20th ACM Symposium on Parallel Algorithms and Architectures*, San Jose, California, June 2011.
7. G. Blelloch, P. Gibbons, and H. Simhadri. Low depth cache-oblivious algorithms. In *Proceedings of the 22nd ACM symposium on Parallelism in algorithms and architectures*, pages 189–199. ACM, 2010.
8. R. D. Blumofe, C. F. Joerg, B. C. Kuszmaul, C. E. Leiserson, K. H. Randall, and Y. Zhou. Cilk: An efficient multithreaded runtime system. Journal of Parallel and Distributed computing, 37(1):55–69, Aug. 1996.
9. R. D. Blumofe. *Executing Multithreaded Programs Efficiently*. PhD thesis, Department of Electrical Engineering and Computer Science, Massachusetts Institute of Technology, Sept. 1995. MIT Laboratory for Computer Science Technical Report MIT/LCS/TR-677.
10. D. Butenhof. Programming with POSIX threads. Addison-Wesley Longman Publishing Co., Inc., Boston, MA, USA, 1997.
11. D. Chase and Y. Lev. Dynamic circular work-stealing deque. In *Proceedings of the seventeenth annual ACM symposium on Parallelism in algorithms and architectures*, page 28. ACM, 2005.
12. S. Chen, P. Gibbons, M. Kozuch, V. Liaskovitis, A. Ailamaki, G. Blelloch, B. Falsafi, L. Fix, N. Hardavellas, T. Mowry, et al. Scheduling threads for constructive cache sharing on CMPs. In *Proceedings of the nineteenth annual ACM symposium on Parallel algorithms and architectures*, pages 105–115. ACM, 2007.
13. Q. Chen, M. Guo, and Z. Huang. Cats: Cache aware task-stealing based on online profiling in multi-socket multi-core architectures. In *the 26th International Conference on Supercomputing*, pages 163–172. IEEE, 2012.
14. Q. Chen, Z. Huang, M. Guo, and J. Zhou. CAB: Cache-aware Bi-tier task-stealing in Multi-socket Multi-core architecture. In *the 40th International Conference on Parallel Processing*, pages 722–732, 2011.
15. Q. Chen, M. Guo, and Z. Huang. Adaptive cache aware bi-tier work-stealing in multi-socket multi-core architectures. IEEE Transactions on Parallel and Distributed Systems, 24(12):2334–2343, 2013.
16. R. Cole and V. Ramachandran. Analysis of Randomized Work Stealing with False Sharing. *ArXiv e-prints*, Mar. 2011.
17. X. Ding, K. Wang, and X. Zhang. ULCC: a user-level facility for optimizing shared cache performance on multicores. In *Proceedings of the ACM SIGPLAN Symposium on Principles and Practice of Parallel Programming*, pages 103–112, 2011.
18. M. Frigo, C. E. Leiserson, H. Prokop, and S. Ramachandran. Cache-oblivious algorithms. In *the 40th Annual Symposium on Foundations of Computer Science*, pages 285–297, New York, USA, 1999. IEEE.

19. A. Gerasoulis and T. Yang. A comparison of clustering heuristics for scheduling directed acyclic graphs on multiprocessors. Journal of Parallel and Distributed Computing, 16(4):276–291, 1992.
20. W. Gropp, E. Lusk, and A. Skjellum. *Using MPI: portable parallel programming with the message passing interface*. MIT Press, 1999.
21. Y. Guo, R. Barik, R. Raman, and V. Sarkar. Work-first and help-first scheduling policies for async-finish task parallelism. In *the 23th IEEE International Parallel and Distributed Processing Symposium*, pages 1–12. IEEE, 2009.
22. Y. Guo, J. Zhao, V. Cave, and V. Sarkar. Slaw: a scalable locality-aware adaptive work–stealing scheduler. In *the 24th IEEE International Parallel and Distributed Processing Symposium*, pages 1–12. IEEE, 2010.
23. D. Hendler and N. Shavit. Non-blocking steal-half work queues. In *Proceedings of the 21th annual symposium on Principles of distributed computing*, pages 280–289. ACM, 2002.
24. D. Hendler, Y. Lev, M. Moir, and N. Shavit. A dynamic-sized nonblocking work stealing deque. *Sun Microsystems, Inc. Technical Reports; Vol. SERIES13103*, page 69, 2005.
25. D. Lea. A Java fork/join framework. In *Proceedings of the ACM 2000 conference on Java Grande*, pages 36–43. ACM, 2000.
26. J. Lee and J. Palsberg. Featherweight X10: a core calculus for async-finish parallelism. In *Proceedings of the 15th ACM SIGPLAN symposium on Principles and practice of parallel computing*, pages 25–36. ACM, 2010.
27. D. Leijen, W. Schulte, and S. Burckhardt. The design of a task parallel library. ACM SIGPLAN Notices, 44(10):227–242, 2009.
28. C. Leiserson. The Cilk++ concurrency platform. In *Proceedings of the 46th Annual Design Automation Conference*, pages 522–527. ACM, 2009.
29. M. M. Michael, M. T. Vechev, and V. A. Saraswat. Idempotent work stealing. In *Proceedings of the 14th ACM SIGPLAN symposium on Principles and practice of parallel programming*, pages 45–54. ACM, 2009.
30. S. L. Olivier, A. K. Porterfield, K. B. Wheeler, and J. F. Prins. Scheduling task parallelism on multi-socket multicore systems. In *Proceedings of the 1st International Workshop on Runtime and Operating Systems for Supercomputers*, pages 49–56, Tucson, Arizona, 2011. ACM.
31. J.-N. Quintin and F. Wagner. Hierarchical work-stealing. In *Proceedings of the 16th international Euro-Par conference on Parallel processing: Part I*, pages 217–229. Springer-Verlag, 2010.
32. J. Reinders. *Intel threading building blocks*. O'Reilly, 2007.
33. R. Van Nieuwpoort, T. Kielmann, and H. E. Bal. Efficient load balancing for wide-area divide-and-conquer applications. In *In ACM SIGPLAN Symposium on Principles and Practice of Parallel Programming*. Citeseer, 2001.
34. L. Wang, H. Cui, Y. Duan, F. Lu, X. Feng, and P. Yew. An adaptive task creation strategy for work-stealing scheduling. In *Proceedings of the 8th annual IEEE/ACM international symposium on Code generation and optimization*, pages 266–277. ACM, 2010.
35. J. Zhang, Z. Huang, W. Chen, Q. Huang, and W. Zheng. Maotai: View-Oriented Parallel Programming on CMT processors. In *37th International Conference on Parallel Processing*, pages 636–643. IEEE, 2008.

Chapter 4
Work-Stealing for NUMA-enabled Architecture

Abstract Modern mainstream powerful computers not only adopt multi-socket multi-core CPU architecture, but also adopt the Non-Uniform Memory Access (NUMA)-based memory architecture. Although the CAB scheduler introduced in Chap. 3 can effectively improve the shared cache utilization, it still leads to severe remote memory accesses in these computers that significantly degrades the performance of memory-bound applications. To solve this problem, in this chapter, we introduce scheduling techniques that can better utilize both the shared cache in CPUs and the NUMA-based memory system.

4.1 Chapter Organization

In this chapter, we first analyze the existing NUMA unawareness problem of the existing random work-stealing schedulers in Sect. 4.2. In Sect. 4.3, we overview the techniques on improving task scheduling for NUMA-based memory system. Specifically in Sect. 4.4, we present the state-of-the-art technique, locality-aware work-stealing (LAWS), that optimize work-stealing for NUMA-based memory system. After that, we evaluate the performance of LAWS in Sect. 4.10.

4.2 Background and Existing Problems

As we present in Fig. 1.2, modern shared-memory MSMC computers often employ NUMA-based (*Non-Uniform Memory Access*) memory system, in which the whole main memory is divided into multiple memory nodes and each node is attached to the socket of a chip. The memory node attached to a socket is called its local memory node and those that are attached to other sockets are called remote memory nodes.

Part of contents in this chapter has been published through ACM Transactions on Architecture and Code Optimization. Reprinted from Ref. [31], with permission from ACM. Figures 4.1, 4.5, 4.7, 4.8 and 4.9 in this chapter have been published through ACM Transactions on Architecture and Code Optimization. Reprinted from Ref. [31], with permission from ACM.

© Springer Nature Singapore Pte Ltd. 2017

Q. Chen and M. Guo, *Task Scheduling for Multi-core and Parallel Architectures*,
https://doi.org/10.1007/978-981-10-6238-4_4

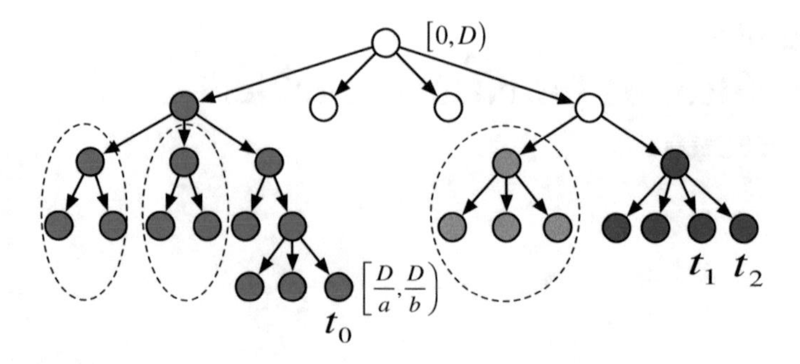

Fig. 4.1 An example task graph for Divide-and-Conquer programs

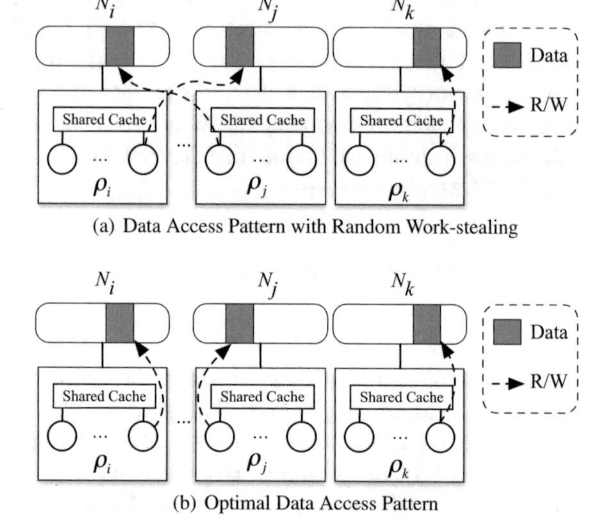

Fig. 4.2 The data access pattern in random work-stealing, and the optimal data access pattern on NUMA-enabled MSMC architectures

The cores of a socket access its local memory node much faster than the remote memory nodes.

On an MSMC architecture with NUMA-base memory system, Fig. 4.2 gives an example task graph for divide-and-conquer programs and the corresponding data access pattern with the traditional random work-stealing. In a divide-and-conquer program, its data set is recursively divided into several parts until each of the leaf tasks only processes a small part of the whole data set.

Suppose the task graph in Fig. 4.1 runs on an MSMC architecture with a NUMA memory system as shown in Fig. 4.2(a). In the NUMA-enabled MSMC architecture, a memory node N_i is attached to the socket ρ_i. In Linux memory management for NUMA, if a chunk of data is first accessed by a task that is running on a core of the socket ρ, a physical page from the local memory node of ρ is automatically allocated to the data. This data allocation strategy employed in Linux kernel and Solaris is called *first touch strategy*.

For a parallel program, its data set is often first accessed by tasks in the first iteration or an independent initialization phase. Taking advantage of the first-touch strategy, if the tasks are scheduled to different sockets, the whole data set of the program that has the task graph in Fig. 4.1 is split and stored in different memory nodes as in Fig. 4.2.

However, traditional random work-stealing suffers from two main problems when scheduling the task graph in Fig. 4.1 in NUMA-enabled MSMC architectures. First, most tasks have to access their data from remote memory nodes in all the iterations. Second, the shared caches are not utilized efficiently.

As for the first problem, suppose the whole data set of the program in Fig. 4.1 is $[0, D)$, and the task t_0 is the first task that accesses the part of the data $[\frac{D}{a}, \frac{D}{b})$ $(a > b \geq 1)$. If task t_0 is scheduled to socket ρ_i, the part of the data $[\frac{D}{a}, \frac{D}{b})$ is automatically allocated to the memory node, N_i, of socket ρ_i, due to the first touch strategy. Suppose task t_0 in a later iteration processes the data $[\frac{D}{a}, \frac{D}{b})$. Due to the randomness of work-stealing, it is very likely that t_0 is scheduled to socket ρ_j instead of ρ_i. In this situation, t_0 is not able to access (read/write) its data from its fast local memory node, instead it has to access a remote memory node for its data. If we can carefully schedule the tasks so that each task is schedule to the socket where the memory node stores its data as shown in Fig. 4.2(b), the data access time can be significantly reduced and the performance of memory-bound applications can be improved in consequence.

As for the second problem, as we discussed in Chap. 3, neighbor tasks (for instance, t_1 and t_2 in Fig. 4.1) are still likely to be scheduled to different sockets due to the randomness of stealing in traditional work-stealing schedulers. This causes more shared cache misses as neighbor tasks in a task graph often share some data. For example, in Fig. 4.1, both t_1 and t_2 need to read all their data from the main memory if they are scheduled to different sockets. However, if t_1 and t_2 are scheduled to the same socket, their shared data is only read into the shared cache once by one task, while the other task can read the data directly from the shared cache.

According to the above analysis, there are three key challenges have to be resolved in an effective task scheduling policy for NUMA-enabled MSMC architectures. First, the policy should be able to balance the workload across multiple sockets. Second, the policy should schedule the tasks carefully, so that each task accesses data from cache as much as possible. Third, if the data of a task is not in the cache, the task should be able to find its data from the local memory node. Only when the above three challenges are resolved, the policy can guarantee load balancing while minimizing the cost of accessing data. In the next section, we discuss prior NUMA-aware task scheduling solutions.

4.3 Prior Solutions

A large amount of prior work has been done to improve the performance [18] of work-stealing policy on various hardwares. Related to LAWS, there are two main approaches for improving the performance of memory-intensive programs in MSMC architectures: *reducing shared cache misses* and *increasing local memory accesses*.

The work of reducing shared cache misses has been discussed earlier in Chap. 3. Besides, a large amount of prior work also improve the performance of a particular application [6, 28, 31] or general applications [24, 25, 30] by increasing local memory accesses in NUMA memory system. For instance, Shaheen et al. [28] proposed a scalable cache aware scheme for iterative stencil computations, nuCATS; and a scalable cache oblivious scheme for iterative stencil computations, nuCORALS. The two schemes improved the performance of iterative stencil computations for NUMA memory system by optimizing temporal blocking and tiling. While the two schemes focused on the tiling scheme for stencil programs, through online scheduling, we aim to improve the performance of iterative stencil programs without changing the tiling scheme.

In the following of this section, we introduce several representative work on improving the performance of memory-intensive applications in MSMC architectures through increasing local memory accesses.

4.3.1 Random Pushing

A well-known variation of random stealing in work-stealing policy is random pushing [33] With random pushing, after a worker generates a new task, the worker checks whether the length of its task queue exceeds a pre-defined threshold. If this is the case, a task from the tail of its task queue is popped out and then pushed into the task queue of a randomly chosen peer worker. The task at the tail of a task queue has the largest workload. This approach aims at minimizing processor idle time because tasks are pushed ahead of time, before they are actually needed, but comes at the expense of additional communication overhead. One might expect random pushing to work well in a wide-area setting, because its communication is asynchronous. Thus, it is less sensitive to high wide-area round-trip times than work stealing. A problem with random pushing, however, is that the algorithm is not stable. Under high work loads, task pushing results in useless overhead, because all the workers already have work.

If random pushing is used in the scenario of distributed cluster, the task will be sent through network, causing bandwidth congestion. Different from traditional work-stealing, random pushing does not adapt its WAN utilization to bandwidth and latency as it lacks a bound for the number of messages that may be sent, i.e. there is no inherent flow-control mechanism.

Another drawback of random pushing is that the memory space is not bounded in random pushing. This is mainly because tasks may be pushed away as fast as they can be generated, and have to be stored at the receiver. In order to avoid exceeding communication buffers, existing implementation of random pushing adds an upper limit of tasks sent by each node that can be in transit simultaneously. This upper limit has to be optimized manually. Additionally, a threshold value must be found that specifies when tasks will be pushed away. A single threshold value is not likely to be optimal for all applications, or not even for one application with different network bandwidths and latencies. In theory, without considering network bandwidth and latency, simply pushing away all the generated tasks is proved to perform well [33]. However, in practice, it has too much communication and marshalling overhead for fine-grained divide-and-conquer applications.

4.3.2 Cluster-Aware Hierarchical Stealing (CHS)

In distributed cluster, traditional random work-stealing and random task pushing both suffer from heavy network communication. In emerging MSMC architecture, they also incurs heavy inter-connect network communication. In order to solve this problem, researchers proposed Cluster-aware Hierarchical Stealing (CHS) for balancing the workload of divide-and-conquer applications in wide-area systems by minimize network communication [33]. The key idea of CHS is to arrange nodes/processors/cores in a tree-shaped topology, and to send steal messages along the edges of the tree. When a node/processor/core is free, it first asks its child nodes/processors for a new task. If the children nodes/processors do not have queued tasks, steal messages will recursively descend the tree. Only when the entire subtree does not have queued tasks, the steal messages will be sent upwards in the tree, asking parent nodes for work.

CHS scheme exhibits much higher data locality, because the tasks inside the same subtree will always complete before the stealing messages are sent to the parent node of the subtree. By arranging the nodes of a cluster in a tree shape, the algorithms locality can be used to minimize network communication. It is also possible that multiple clusters work together to process an parallel application. When the root node of a cluster finds its own cluster to be idle, it sends a steal message to the root node of another, randomly selected cluster. Figure 4.3 shows an example arrangement of the clusters. In the figure, each small filled square is a computer node; each large square is a cluster. Different clusters communicated with each other using their root nodes.

Due to the tree-shaped topology, CHS has two drawbacks. First, the root node of a cluster waits until the entire cluster becomes idle before starting cross-cluster steal attempts. During the round-trip time of the steal message, the entire cluster remains idle. Second, the preference for stealing further down the tree results in jobs with finer granularity to be stolen first, leading to high network communication overhead, due to tremendous task transfers.

4.3.3 Cluster-Aware Load-Based Stealing (CLS)

As we discussed earlier, both random pushing and CHS suffers from high network communication overhead, especially for application with fine-grained tasks. Espe-

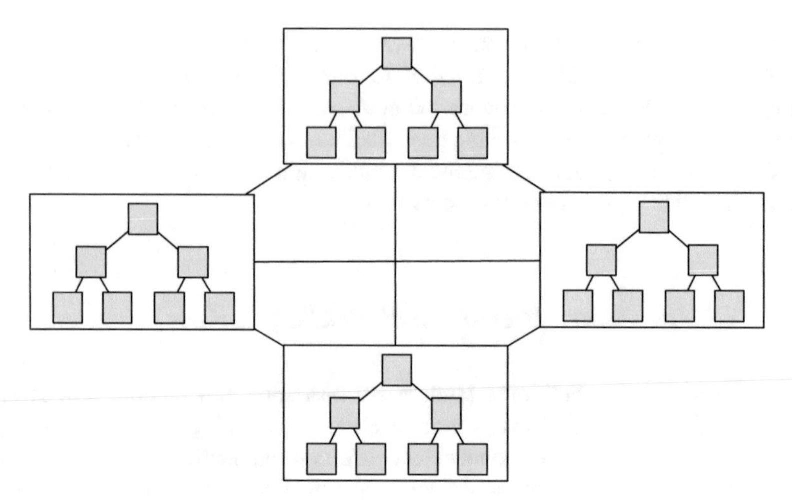

Fig. 4.3 An example arrangement of the clusters. Nodes are connected in tree shapes and multiple trees are connected via wide-area links

cially, when stealing a task from a remote cluster, the whole cluster stalls, waiting for the new task. The high network communication overhead and the cluster stalls together result in the poor performance of random pushing and CHS. In order to address the two problems, Cluster-aware Load-based Stealing (CLS) is proposed [33]. The key idea behind CLS is to combine random stealing inside clusters with cross-cluster task prefetching performed by one coordinator node per cluster. With a single cluster, CLS is identical to random work-stealing.

Insider a cluster, each node adopts traditional random work-stealing to steal tasks from each other. Because there is no preference for stealing fine-grained tasks, it reduces the network communication overhead compared with the tree-based approach in CHS. Between different clusters, in order to minimize the cross-cluster communication, CLS only allows coordinator nodes to perform cross-cluster work stealing. Furthermore, in order to avoid cluster stalling, CLS allows the coordinator nodes to prefetch tasks. The prefetching operation requires a careful balance between communication overhead and processor idle time. On the one hand, if the tasks are prefetched too early, the communication overhead grows unnecessarily. On the other hand, if the tasks are prefetched too late, processors may become idle. Pure work-stealing can be seen as one extreme of this tradeoff, where jobs are never prefetched. Pure work-pushing is the other extreme where jobs are always transferred ahead of time.

In CLS, prefetching is controlled by the workload information from the compute nodes. Each node periodically sends its workload information to the cluster coordinator that monitors the overall workload of its cluster. The compute nodes send their load messages asynchronously, keeping the overhead small. Furthermore, when a node becomes idle, it immediately sends a load message (with the value zero) to the coordinator. A good interval for sending the load messages is subject to parameter tuning. When the total cluster load drops below a specified threshold value, the coordinator initiates cross-cluster steal attempts to randomly chosen nodes in remote clusters. The coordinator can hide the high cross-cluster round-trip transfer time by overlapping communication and computation, because the cross-cluster prefetch messages are asynchronous.

Another tunable parameter is the value of the threshold that is used to trigger the cross-cluster prefetching. This parameter strongly depends on the application granularity and the network latency and bandwidth. An option is to use the length of the work queue as load indicator. This indicator might not be very accurate, due to the job granularity that descends with increasing recursion depth. It is still an open problem to find the most effective load indicator, although prior work [21] found that the queue length on a node is a relative effective way to indicate the workload on the node.

4.3.4 Cluster-Aware Random Stealing (CRS)

The cluster-aware load-based stealing scheme has two main drawbacks. First, according to the above discussion, only when the parameters (e.g., threshold used to trigger the cross-cluster prefetching, idle interval) are carefully tuned, CLS is able to minimize both intra-cluster and cross-cluster network communication overheads. Second, the prefetched tasks are stored on a centralized coordinator node rather than on the idle nodes themselves. In this case, with the high cross-cluster round-trip time, the coordinator nodes might become a stealing bottleneck if the local nodes together compute tasks faster than they can be prefetched by the coordinator node.

Cluster-aware Random Stealing (CRS) [33] is proposed to overcome the above two problems. Same to CLS, it uses random task stealing inside clusters. Different from CLS, CRS scheme uses a different approach to perform cross-cluster stealing. The key idea behind CRS is to omit centralized coordinator nodes. Instead, a decentralized control mechanism is proposed and adopted in CRS for the cross-cluster communication directly in the worker nodes. In more detail, Algorithm 4 gives the algorithm of CRS.

As observed from the algorithm, each node can directly steal tasks from nodes in remote clusters, but at most one task at a time. Whenever a node becomes idle, it first attempts to steal from a node in a remote cluster. This cross-cluster steal request is sent asynchronously. Instead of waiting for the result, the thief simply sets a flag and performs additional, synchronous steal requests to nodes within its own cluster, until it finds a new task. When a local task is found, the cross-cluster steal request is not cancelled. As long as the flag is set, only local stealing will be performed. The handler routine for the cross-cluster reply simply resets the flag and, if the request was successful, puts the new task into the task queue.

The CRS scheme also implements an efficient way of task prefetching that delivers the new task directly on the idle node and does not need parameter tuning. The implication of this scheme is that many remote clusters will be asked for tasks concurrently when a large part of a cluster is idle. As soon as one remote steal attempt

Algorithm 4 Pseudo code for Cluster-aware Random Stealing.

```
void cluster_aware_random_stealing(void) {
  while(NOT exiting) {
    task = queue_get_from_head();
    if(task) {
      execute(task);
  } else {
    if(nr_clusters > 1 AND NOT stealing_remotely) { /* no wide-area message in transit */
      stealing_remotely = true;
      send_async_steal_request(remote_victim());
    }
    task = send_steal_request(local_victim());
    if(task) queue_add_to_tail(task);
    }
  }
}
void handle_cross_cluster_reply(Task task) {
  if(task) queue_add_to_tail(task);
  stealing_remotely = false;
}
```

is fulfilled, the work will be quickly distributed over the whole cluster, because local steal attempts are performed during the long cross-cluster round-trip time. Therefore, when tasks are found in a remote cluster, the work is propagated quickly.

Compared to traditional random work-stealing, CRS significantly reduces the number of messages sent between different clusters. CRS has the advantages of random stealing, but hides the long cross-cluster round-trip time by additional, local stealing. The first task to arrive will be executed. No extra load messages are needed, and no parameters have to be tuned.

The above four schemes (i.e., random pushing, CHS, CLS, CRS) are proposed to improve the performance of parallel applications on large-scale distributed clusters. Table 4.1 [33] compares the four schemes. Observed from the table, the CRS scheme performs the best compared with the other three schemes. These schemes are also able to be adapted to NUMA-enabled MSMC architectures where a socket is treated to be a cluster and the cores in the same socket are treated to be the nodes in the same cluster.

4.3.5 TATL

Besides the above four schemes, Vikranth et al. [30] proposed a dynamic topology-aware work-stealing scheme for on-chip NUMA-enabled multi-core processors based on the topology of underlying hardware.

In TATL, if there are *M* sockets, and there are N cores per socket, then *M* worker pools are created where each pool contains *N* worker threads there by grouping the

Table 4.1 Comparison between random pushing, CHS, CLS, and CRS

Scheme	Optimization goal	Drawbacks	Task transfer
Random pushing	Idle time	Unstable heavy cross-cluster comm.	Synchronous
CHS	Cross-cluster comm.	Cluster stalling heavy inter-cluster comm.	synchronous
CLS	Inter-cluster comm. Cross-cluster comm.	Slow task distribution prefetching bottleneck	Synchronous in cluster Prefetching across cluster
CRS	Inter-cluster comm. Cross-cluster comm.	/	Synchronous in cluster prefetching across cluster

total worker threads into M stealing domains. By restricting the task stealing within the same domain, the number of cross chip references and remote cache misses are reduced. Task stealing from a remote domain is allowed only when a thief worker is unable to find a victim worker in its local domain. Grouping $M \times N$ worker threads in M domains of N workers each gives the advantage of flexible implementation and does not cause any overhead. The stealing domains also allow the runtime to be easily scalable.

TATL also updates the random victim selection policy in traditional work-stealing. In TATL scheme, each worker is responsible for advertising itself whenever its queue length reaches the threshold value. Equations 4.1 and 4.2 computes the minimum threshold value and the maximum threshold value for the queue size respectively. In the two equations, C is the capacity of the task queue; T_{push} and T_{pop} denote the time taken to perform push in or pop out to the double-ended task queue; λ and μ represent the arrival and processing rates of task queue respectively.

$$S = C - \lambda T_{push} + \mu T_{pop} \tag{4.1}$$

$$s = \mu T_{pop} \tag{4.2}$$

Leveraging the two thresholds, if the queue size of a worker is smaller than S, the run-queue of the worker thread enters into THIEF state. Otherwise, if the queue size of a worker is larger than s, the run-queue of the worker thread enters into VICTIM state. When a worker in THIEF state completes all its tasks, the thief worker first searches the list of workers whose state is already VICTIM in the same stealing domain. In emerging implementation of TATL, a status bit is added to each worker queue to represent either THIEF or VICTIM. The thief worker searches only the run-queues with status VICTIM. This solves the problem of randomly choosing the victim and failure to find a queue with enough number of tasks. When a thief worker tries to search for a victim worker, it can find the victim easily by looking at the bit. Hence the delays involved in repeated attempts are removed.

Algorithm 5 Algorithm of stealing tasks in TATL

```
if ( localTaskQueue.size == THRESHOLD_MAX_SIZE )
then
  this.status = VICTIM;
endif
if ( ! isEmpty(localTaskQueue) ) then
  run:
    popAtFront(&localTaskQueue, &task);
    execute task;
    if (localTaskQueue.size == THRESHOLD_MIN_SIZE ) then
      this.status = THIEF;
    endif
else
  this.status = THIEF;
end if
taskQueue= searchForVictimQueue ( thisStealingDomain );
popAtRear(&taskQueue, &task);
if ( task ) then
  pushAtRear(&localTaskQueue , task );
  goto run;
else
  runQueue= searchForVictimQueue ( remoteStealingDomain);
  popAtRear(&taskQueue, &task);
  if ( task ) then
    pushAtRear(&localTaskQueue , task );
    goto run;
  endif
end if
```

Algorithm 5 shows the algorithm of stealing tasks in TATL. In the algorithm, the function call *searchForVictimQueue()* searches for the run queues whose status is already set to VICTIM. The values of THRESHOLD_MAX_SIZE and THRESHOLD_MIN_SIZE are computer using the S and s variables from Eqs. 4.1 and 4.2.

4.3.6 NUMALB

NUMALB [25], a NUMA-aware load balancer, is proposed to improve parallel system performance based on Charm++ [19], a parallel runtime system that provides an object oriented parallel programming language.

In most cases, once the data (e.g. a message) is touched, this memory policy will not perform any data migration to enhance memory affinity. This might result in sub-optimal data placement on NUMA platforms. For instance, we can imagine a situation where some messages have been generated and originally allocated on core 0 of NUMA node 0. After that, these messages are sent to core 1 of NUMA node 1 and after several hops they end up on core N of NUMA node N. All message sends are pointer exchanges of data that were originally allocated and touched in the memory of core 0. In such a scenario, several remote accesses will be generated for every communication.

NUMALB balances the workload while avoiding unnecessary migrations and reducing cross-core communication. Generally speaking, based on runtime information collected online, NUMALB adopts a heuristic algorithm to reduce the load imbalance of parallel applications in NUMA-enabled MSMC architecture.

4.3.6.1 Obtaining Runtime Information

NUMALB collects two types of runtime information: *application data* and *NUMA topology*. *Application data* comprises all information about the paral- lel application that can be probed at runtime: task execution times, communication information, and the assignment chosen by the scheduler at a given time. In CHARM++ RTS, this information can be dynamically obtained during the execution of the application. The *NUMA topology* comprises all information that can be gathered at runtime about the machine hardware that is running the application. A NUMA computer can be characterized in terms of the number of NUMA nodes, cache memory sizes, sharing of cache hierarchies among cores and grouping of NUMA nodes.

NUMALB defines a **NUMA factor** to synthesize both the NUMA topology and the memory access penalties (it is more expensive to read data from remote memory node than local memory node). The *NUMA factor* represents the overhead to access remote data and is defined as in Eq. 4.3:

$$F(i, j) = \frac{L_{ij}}{L_{ii}} \tag{4.3}$$

In the equation, $F(i, j)$ is the NUMA factor of accessing data from cores in socket i to the data stored in the NUMA node j. L_{ij} is the data read latency from cores in socket i to NUMA node j; and L_{ii} is the data read latency from the local memory node of socket i. This factor is then computed for all NUMA nodes of the target machine, resulting in a square matrix of NUMA factors. Thus, the main advantages of using the NUMA factor as a topology indicator is that it is generic (can be easily computed for different NUMA machines) and aggregates the differentiating features of NUMA machines. In addition, the NUMA factor can be precomputed, which reduces the overhead of using it.

4.3.6.2 Load Balancing Heuristic

It is not possible to compute an assignment of tasks on to available processors that optimally equalizes the load in polynomial time (unless $P = NP$). Moreover, in the general case, a good scheduler should not make any assumptions about the application that will be executed, so it is also impossible to use precomputed assignments instead of online scheduling. Thus, in practice, in order to compute a good (approximated) assignment in a reasonable amount of time, NUMALB employs a heuristic algorithm online.

The heuristic in NUMALB works like a classical List Scheduling algorithm [33], where tasks are rescheduled from a priority list and assigned to less loaded processors in a greedy manner. List schedule algorithms usually are fast to compute and provide good results in practice. The main idea of the heuristic is to improve application performance by mapping tasks to cores while reducing the costs of unbalanced computation and remote communications. The heuristic is based on the cost function defined in Eq. 4.4 for mapping of a task t on to core p.

$$cost(t, p) = load(p) + \alpha \times (r_{comm}(t, p) \times F(comm(t), node(p)) - l_{comm}(t, p)) \tag{4.4}$$

In the equation, $load(p)$ represents the total load of core p, l_{comm} represents the number of messages sent from task t to the tasks on cores of the same socket as core p, and r_{comm} expresses the number of messages sent from task t to the tasks on other NUMA nodes and is multiplied by the NUMA factor between the NUMA node of core p ($node(p)$) and the NUMA nodes where these communicating tasks are mapped ($comm(c)$). Finally, α controls the weight that the communication costs have over the execution time. The heuristic uses the number of exchanged messages because it represents the amount of accesses to the shared memory. Since messaging time is related to the access latency, the cost is multiplied by the NUMA factor when considering remote accesses. In addition, local communications are subtracted from the overall cost to favor their occurrence.

Leveraging the heuristic described above and list scheduling, NUMALB always picks the heaviest (largest execution time) unassigned task and assigns it to the core that presents the smaller cost until all the tasks are allocated to appropriate cores.

However, NUMALB is only able to increase local memory accesses, it is not able to reduce shared cache misses. Furthermore, it makes decision relying on the number of messages sent from a task to the other tasks. This information is hard to obtain at runtime online.

4.3.7 *Offline Technique for Unstructured Parallelism*

The techniques we mentioned above mainly target applications with structured parallelism. They are not able to exploiting locality for applications with unstructured parallelism due to the following reasons. First, the lack of dependency information implies the scheduler must obtain additional information from the workload to synthesize locality structure. Without understanding what the crucial information is, runtime and storage overheads for collecting the information can be significant. Second, the larger degrees of freedom in scheduling increases algorithmic complexity. Having many degrees of freedom implies many grouping and ordering choices, and enumerating all combinations is prohibitive. Third, the complexity of many-core cache hierarchies makes the process all the more complicated. Grouping and ordering decisions must optimize locality across all cache levels, whether the hierarchy being shared or private.

In order to exploit locality for applications with unstructured parallelism, based on METIS [20], Yoo et al. [32] proposed an offline graph-based locality analysis framework to analyze the inherent locality patterns of workloads. Leveraging the analysis results, tasks are grouped and mapped according to cache hierarchy through recursive scheduling. We introduce the two steps as follows.

4.3.7.1 Graph-Based Locality Analysis

A locality-aware schedule should map tasks to cores, taking into account both locality and load balance. Executing a set of tasks (a task group) on cores that share one or more levels of cache captures data reuse across tasks. Similarly, executing tasks in an optimal order minimizes the reuse distance of shared data between tasks, which makes it easier for caches to capture the temporal locality. Generating a locality-aware schedule depends on understanding how task groups should be formed, and when ordering will matter.

In a schedule, both *task grouping* and *task ordering* affect the performance of a parallel application. Executing a set of tasks (a task group) on cores that share one or more levels of cache captures data reuse across tasks. Similarly, executing tasks in an optimal order minimizes the reuse distance of shared data between tasks, which makes it easier for caches to capture the temporal locality.

The graph-based locality analysis framework proceeds as follows: (1) each workload is profiled to collect the data access traces at cache line granularity, and discard ordering information to obtain read and write sets for each task. (2) Using the set information, the framework constructs a task sharing graph. In a task sharing graph $G(V, E)$, a vertex represents a task, and an edge denotes sharing. A vertex weight is the task size in terms of number of dynamic instructions, and an edge weight is the number of cache lines shared between the two tasks connected by the edge. (3) the framework then partitions the graph to form task groups, and observes some metrics to determine the "right" task group size and the impact of ordering.

Even with profile information about each tasks read and write sets, creating an "optimal" set of task groups is still NP-hard. To solve this problem, a heuristic graph partitioning tool, METIS [20], is adopted to generate quality task groups. METIS divides the vertices from a task sharing graph into a given number of groups, while trying to (a) maximize the sum of edge weights internal to each group (i.e., data sharing captured by a task group), and (b) equalize the sum of vertex weights in each group (i.e., balance load).

4.3.7.2 Recursive Task Scheduling

Based on the above analysis results, computation is able to be mapped onto an actual cache hierarchy appropriately. Specifically, recursive scheduling, which (1) matches

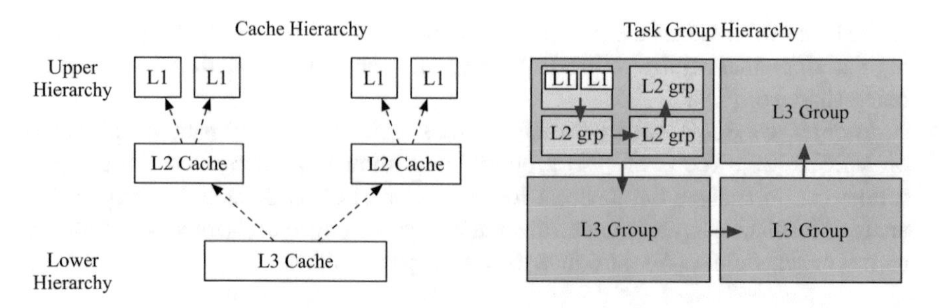

Fig. 4.4 Generating recursive task groups [32]. Different levels of groups are sized to fit in a particular cache level. Colored arrows denote the group order determined over task groups

task group working sets and (2) applies ordering across all cache levels can be adopted in this scenario.

Creating an optimal order for tasks is NP-hard. Heuristic-based techniques can be applied in this scenario to provide high quality ordering. For instance, Prim's algorithm can be applied to construct a maximum spanning tree (MST), and use the order that the vertices are added to the MST. In architectural terms, Prims algorithm accumulates the read and write sets of scheduled tasks, and picks the task whose read and write sets exhibit the maximum intersection with the cumulative sets as the next task to execute. After that, in order to construct a task order, the MST is applied on a task sharing graph. In order to construct a group order, the task sharing graph is first mapped to a *task group sharing graph*, where each *uber node* represents a task group; then MST is applied to the task group sharing graph.

Under recursive scheduling, in order to maximize the utility of every cache level, tasks are grouped from the bottommost: We can first group tasks so that the task groups working set fits in the last-level cache, and apply ordering over those groups. We can then recursively apply this approach to each of the task groups, targeting one level up in the cache hierarchy each time. For instance, Fig. 4.4 gives an example that groups tasks for an architecture with three levels of caches. In the figure, we first perform grouping on the full set of tasks to create L3 groups, each of which matches the L3 size. Next, we order the L3 groups. For each L3 group, we then decompose it into tasks and create L2 groups to match the L2 size. Then we order the L2 groups. We proceed in this fashion until we finally generate L1 groups, order them, and order their component tasks.

Generating a schedule as shown in Fig. 4.4 results in a hierarchy of task groups. Moreover, since each task group also denotes a scheduling granularity, all the tasks in a group will be executed consecutively. Therefore, a task group will stay resident in its target cache from beginning to end. The existence of a hierarchy among these task groups guarantees that all the groups containing a given task stay resident at their corresponding level in the cache hierarchy, thus exploiting locality across all cache levels.

Because the framework relied on offline analysis, a program has to be executed at least one time before it can achieve good performance in the framework. On the contrary, the technique proposed in this section, LAWS, can improve the performance of programs online without any prerequisite offline analysis, because it can pack tasks into CF subtrees based on online collected information and auto-tuning.

4.4 Design of Locality-Aware Work-Stealing

As we discussed in Sect. 4.3, emerging work-stealing schedulers either are not able to improve shared cache utilization or not able to reduce the remote memory accesses. In this section, we introduce a *Locality-Aware Work-Stealing* (LAWS) scheduler. According to the above analysis, in order to optimize the performance of memory-bound applications, when we design an efficient task scheduler for NUMA-enabled MSMC architecture, the scheduler should obey the following several key guidelines.

- LAWS should have the ability to balance the workload across the sockets/cores.
- LAWS should have the ability to improve the shared cache utilization.
- LAWS should have the ability to reduce the remote memory accesses.

Following the above three key guidelines, LAWS ensures that the workload is balanced and most tasks can access data from either the shared cache or the local memory node. The performance of memory-bound programs can be improved due to balanced workload and shorter data access latency.

More specifically, LAWS consists of a load-balanced task allocator, a cache-friendly task graph partitioner and a triple-level work-stealing scheduler. The load-balanced task allocator can evenly distribute the data set of a parallel program to all the memory nodes and allocate a task to the socket where the local memory node stores its data. The cache-friendly task graph partitioner can divide the task graph of a program into *Cache-Friendly Subtrees* (CF subtrees) so that the shared cache of each socket can be utilized effectively. The CF subtrees is similar to the intra-socket subtrees defined in Chap. 3. Because the purpose of dividing the task graph into subtrees is to improve the cache utilization, we rename the intra-socket subtree to CF subtree in this Chapter. The triple-level work-stealing scheduler is able to schedule all the tasks accordingly to balance the workload and reduce shared cache misses.

Figure 4.5 illustrates the processing flow of an iterative program in LAWS. As shown in Fig. 4.5, in every iteration, the load-balanced task allocator carefully assigns tasks to different sockets to evenly distribute the data set of the program to all the memory nodes and allocate each task to the socket where the local memory node stores its data. In this situation, the workload of different sockets is balanced in general since the time for processing the same amount of data is similar among tasks in well-designed parallel programs. There may be some slight load-unbalance which could be resolved by the triple-level work-stealing scheduler.

After the tasks are allocated to appropriate sockets with the load-balanced task allocator, each socket will still have to execute a large number of tasks. The data involved in these tasks are often too large to fit into the shared cache of a socket.

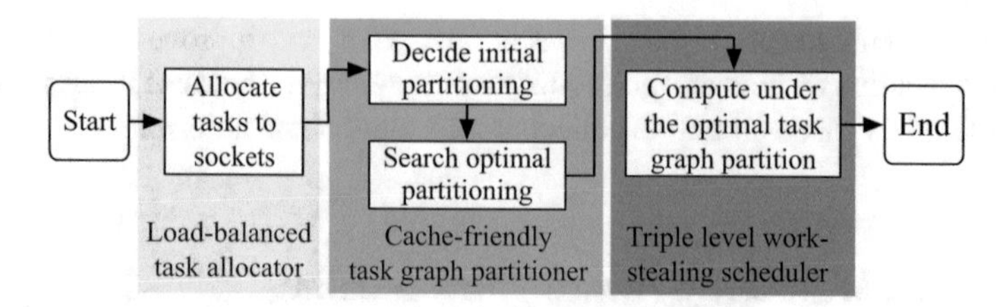

Fig. 4.5 The processing flow of an iterative program in LAWS

As we discussed in Chap. 3, the large data set often results in the TRICI problem in multi-socket multi-core architecture. To utilize the shared cache efficiently, LAWS further leverages the cache-friendly task graph partitioner to packs the tasks allocated to each socket into a large number of CF subtrees that will be executed sequentially based on runtime information collected in the first iteration. Because tasks in the same CF subtree often share some data, the shared data is only read into the shared cache once but can be accessed by all the tasks of the same CF subtree. In this way, the shared cache can be better utilized.

For instance, in Fig. 4.1, the tasks filled with the same color consist of the part of the whole task graph allocated to a socket, and the subtree in each ellipse is a CF subtree. In the first several iterations, the partitioner automatically adjusts the partition of task graph to search for the optimal one that results in the minimum makespan. Because the task graphs of different iterations are the same and the tasks in the same position of the task graphs work on the same part of the data set in an iterative program, the optimal partition for the completed iterations is also optimal for future iterations. Once the optimal partition is found, LAWS partitions the task graph in all the following iterations in a way suggested by the optimal partition.

LAWS adopts a triple-level work-stealing scheduler to schedule tasks in each iteration. The tasks in the same CF subtrees are scheduled within the same socket. If a socket completes all its CF subtrees, it steals a CF subtree from a randomly-chosen socket in order to resolve the possible slight load-unbalance from the task allocator.

4.5 Load-Balanced Task Allocator

The purpose of the load-balanced task allocator is to balance the workload across all the available sockets, and make sure most of the tasks can find their data from the local memory nodes. In LAWS, the load-balanced task allocator is proposed based on an assumption that a task divides its data set into several parts evenly according to its branching degree. This assumption is true in emerging well-designed divide-and-conquer parallel programs.

In order to achieve the above design purpose, the load-balanced task allocator should satisfy two main constraints when allocating tasks to sockets. First, to balance workload, the size of data processed by tasks allocated to each socket should be same

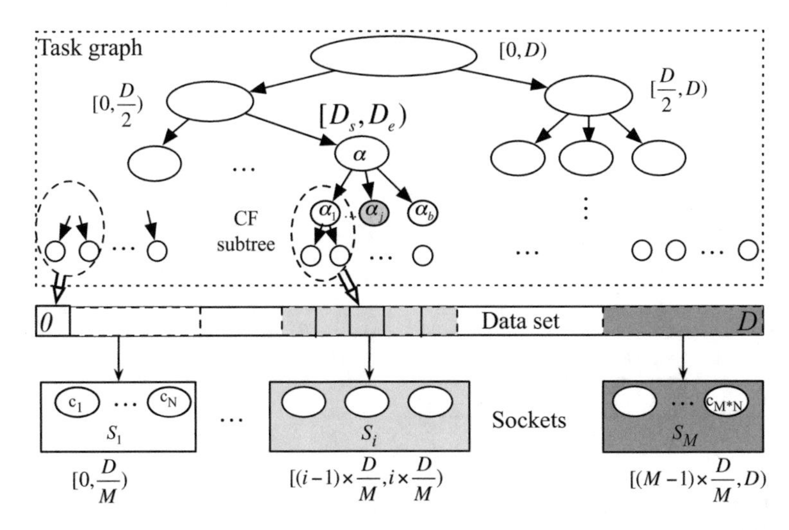

Fig. 4.6 Allocate the tasks of program p to the sockets of an M-socket NUMA-enabled architecture in LAWS

in every iteration. Second, to reduce shared cache misses, the adjacent data should be stored in the same memory node since adjacent data is processed by neighbor tasks that should be scheduled to the same socket. Traditional random work-stealing schedulers do not satisfy the two constraints due to the randomness of stealing. And the CAB scheduler discussed in Chap. 3 only satisfies the second constraint and does not satisfy the first constraint.

We first model how the whole dataset of a parallel program is distributed to the tasks. Suppose a program p runs on an M-socket architecture. If its data set is D, to balance workload, the tasks allocated to each socket need to process $\frac{1}{M}$ of the whole data set. Note that, in the load-balanced task allocator, we do not need to know the real value of D. We use D to represent the whole data set for easy of description. Without loss of generality, as shown in Fig. 4.6, LAWS makes sure that the tasks allocated to the i-th ($1 \leq i \leq M$) socket should process the part of the whole data set ranging from $(i-1) \times \frac{D}{M}$ to $i \times \frac{D}{M}$ (denoted by $[(i-1) \times \frac{D}{M}, i \times \frac{D}{M})$).

To achieve the above objective, LAWS needs to find out each task processes which part of the whole data set. In LAWS, when a task is spawned, the part of its data set is automatically calculated based on the structure of dynamically generated task graph. Take task α that has b sub-tasks and processes the part of data set ranging from D_s to D_e (denoted by $[D_s, D_e)$) in Fig. 4.6 as an example. When its j-th subtask α_j is spawned, the part of the data set that α_j will process can be calculated by Eq. 4.5.

$$[(j-1) \times \frac{D_e - D_s}{b} + D_s, j \times \frac{D_e - D_s}{b} + D_s) \tag{4.5}$$

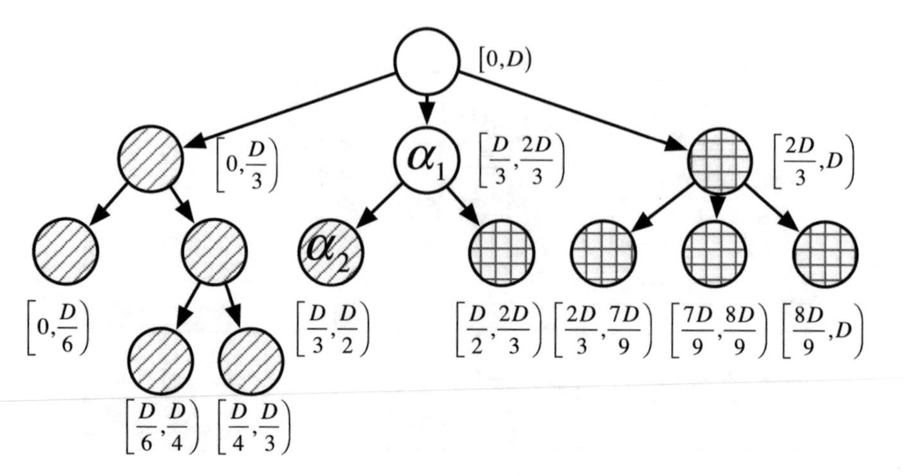

Fig. 4.7 An example of allocating the tasks to the two sockets of a dual-socket NUMA-enabled architecture

When we implement LAWS, all the parameters (i.e., b and j) in Eq. 4.5 are obtained automatically. LAWS records the branching degree of each task at compile time by analyzing task generating pattern and calculates j in Eq. 4.5 when α_j is generated.

In more detail, Fig. 4.7 gives an example of allocating the tasks to the two sockets of a dual-socket architecture. The range of data beside each task is calculated according to Eq. 4.5. In the dual-socket architecture, the tasks that process the data set $[0, \frac{D}{2})$ and $[\frac{D}{2}, D)$ should be allocated to the first socket and the second socket respectively. For instance, in Fig. 4.7, because α_2 is responsible for processing data range $[\frac{D}{3}, \frac{D}{2})$ that is within $[0, \frac{D}{2})$, it should be allocated to the first socket. Due to the same reason, the slash-shaded tasks are allocated to the first socket and the mesh-shaded tasks are allocated to the second socket. If a task is allocated to a socket, all its child tasks are allocated to the same socket. For example, all the tasks rooted with α_2 will be allocated to the first socket.

LAWS records the part of data set stored in the memory node of each socket. By comparing the data set processed by a task with the range of data set stored in each socket, the task can be allocated to appropriate socket. Because the task allocator allocates a task according to the range of its data set, in the following iterations, the tasks processing the same part of the whole data set will be allocated to the same socket. In this way, the tasks in all the iterations can find their data in the local memory node, and the adjacent data is stored in the same memory node. Therefore, the two constraints in designing the load-balanced task allocator are satisfied and the first problem discussed in Sect. 4.2 in all the prior work-stealing schedulers will be solved.

4.6 Cache-Friendly Task Graph Partitioner

On an M-socket architecture, the load-balanced task allocator divides the whole task graph of a parallel program into M parts and allocates the M parts to the M sockets. However, when the input data of the parallel program is large, each socket will still have to execute a large number of tasks. In this case, the data involved in these tasks are often too large to fit into the shared cache of a socket. To utilize the shared cache efficiently, LAWS further divides the task graph allocated to each socket into multiple CF subtrees that will be executed sequentially.

It is worth noting that the work-stealing scheduler and each task often generate some intermediate data during the execution of a program. Therefore, the precise size of data involved in each task is not known during the execution of a parallel program, even if the size of the whole input data of the program is known. It is not trivial to further divide the task graph into CF subtrees lacking of precise data usage information. In order to solve this problem, as described below, we use an auto-running approach based on online-collected profiling information to search for the optimal task graph partitioning.

The following technique of partitioning task graph into CF subtrees only works for iterative programs because it relies on the processing time of different iterations to identify the optimal partition. For non-iterative programs, the compiling-based technique proposed in Sect. 3.4.2.1 can be used to partition the task graph into CF subtrees.

4.6.1 Decide the Initial Partitioning

Similar to LAB task scheduler presented in Chap. 3, the task graph partitioner makes sure that the data accessed by all the intra-socket tasks in each CF subtree can be fully stored into the shared cache of a socket. Note the tasks in the same CF subtree (called *intra-socket tasks*) are scheduled in the same socket and the root task of a CF subtree is called a *CF root task*. In this way, the data shared by tasks in the same CF subtree is read into the shared cache once but can be shared and accessed by all the tasks.

To achieve the above objective, we need to know the size of shared cache used by each task, which cannot be collected directly. Recall that we have introduced the profiling-based technique that can be used to calculated this parameter in Sect. 3.4.2.2. In the cache-friendly task graph partitioner, we leverage the same technique to calculate this parameter. In other words, to circumvent this problem, in the first iteration, for any task α, LAWS collects the number of last level private cache (e.g. L2) misses caused by it. The size of shared cache used by α can be estimated as the number of the above cache misses times the cache line size (e.g., 64 bytes).

For easing of description, we describe the technique here briefly again. For task α, we calculate its *SOID*, which represents the *Size Of Input Data used by all the tasks in the subtree rooted with* α. SOID of α is calculated in the bottom-up manner. Suppose α has m direct child tasks $\alpha_1, ..., \alpha_m$ and their SOIDs are $S_1, ..., S_m$ respectively. SOID of α (denoted by S_α) can be calculated in Eq. 4.6, where M_α equals to the number of last level cache misses caused by α itself times the cache line size.

$$S_\alpha = M_\alpha + \sum_{i=1}^{m} S_i \tag{4.6}$$

Once all the tasks in the first iteration complete, SOIDs of all the tasks are calculated using Eq. 4.6. Based on SOIDs of all the tasks, the cache-friendly task graph partitioner can further divide the part of the task graph allocated to each socket into CF subtrees by identifying all the CF root tasks (known as *leaf inter-socket tasks* in Chap. 3) as follows.

Let S_c represent the shared cache size of a socket. Suppose α's parent task is β, and their SOIDs are S_α and S_β respectively. Then, if $S_\alpha \leq S_c$ and $S_\beta > S_c$, α is a CF root task, which means all the data involved in the descendent tasks of α just fit into the shared cache. If $S_\beta < S_c$, α is an intra-socket task. Once all the CF root tasks are identified, the initial task graph partition is determined.

4.6.2 Search for the Optimal Partitioning

If S_α in Eq. 4.6 precisely equals to the real size of shared cache used by the subtree rooted with α, the data involved in any CF subtree would not exceed the capacity of a socket's shared cache.

However, S_α is only a close approximation due to the following reasons. Suppose tasks α_1 and α_2 in the subtree rooted with α share some data. Although they are allocated to the same socket by the load-balanced task allocator, they can be executed by different cores. In this case, both α_1 and α_2 need to read the shared data to the last level private cache and thus the size of the shared data is accumulated twice in Eq. 4.6. On the other hand, if some data stored in the shared cache has already been pre-fetched into the private cache before, it does not incur last level private cache misses and the size of the pre-fetched data is missed in Eq. 4.6. The multiple accumulation of shared data and the pre-fetching make S_α of Eq. 4.6 slightly larger or smaller than the actual size of shared cache used by the subtree rooted with α.

Due to the approximation in calculating the SOID of each task, the initial task graph partitioning found in Sect. 4.6.1 is only a near optimal task graph partitioning. LAWS further uses an auto-tuning approach to search for the optimal task graph partitioning. In the approach, the task graph partitioner divides the task graph into CF subtrees differently in different iterations, records the execution time of each iteration, and chooses the partition that results the shortest processing time as the optimal partition.

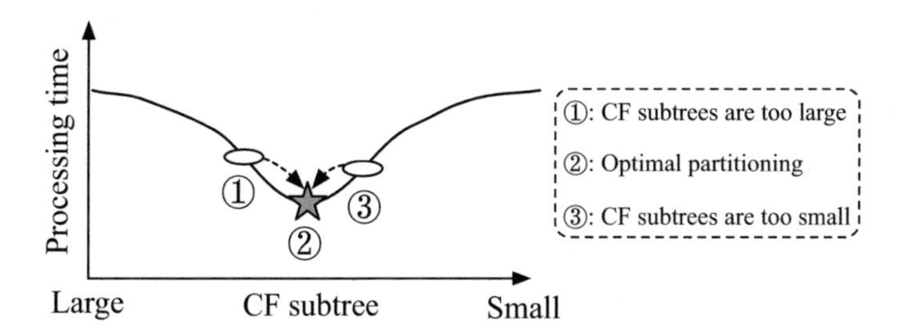

Fig. 4.8 Processing time of an iteration in a parallel program when the task graph is partitioned into subtrees of different sizes

Figure 4.8 shows the execution time of an iteration when the task graph is partitioned differently. If CF subtrees are too large (contain too many intra-socket tasks, point 1 in Fig. 4.8), the data accessed by tasks in each CF subtree cannot be fully stored in the shared cache of a socket. On the other hand, if CF subtrees are too small (contain too few intra-socket tasks, point 3 in Fig. 4.8), the data accessed by tasks in each CF subtree is too small to fully utilize the shared cache.

Because we have found a good start point of the task graph partitioning in the previous step, starting from the initial partitioning of the task graph into CF subtrees found in Sect. 4.6.1, the task graph partitioner evaluates how the different partitioning performs. In more detail, LAWS first evaluates how the smaller CF subtrees perform. If smaller CF subtrees result in shorter processing time of an iteration (point 1 in Fig. 4.8), CF subtrees in the initial partitioning are too large. In this case, the task graph partitioner evaluates the partition that has smaller and smaller CF subtrees until the partitioning that results in the shortest processing time (point 2 in Fig. 4.8) is found. If smaller CF subtrees result in longer processing time of an iteration (point 3 in Fig. 4.8), CF subtrees in the initial partitioning are too small. In this case, the task graph partitioner evaluates the partition that has larger and larger CF subtrees instead until the optimal partitioning is found.

Algorithm 6 gives the auto-tuning algorithm for searching the optimal way to partition the task graph allocated to a socket into smaller CF subtrees. To generate larger or smaller CF subtrees, we select the parent tasks or child tasks of the current CF root tasks as the new CF root tasks.

Since the initial partitioning is already near-optimal, LAWS can find the optimal task graph partition in a small number of iterations. Theoretically, it has a small possibility that some CF subtrees are too large while some other CF subtrees are too small. However, since there are a great many CF subtrees in a task graph, it is too complex to tune the size of every CF subtrees independently in a small number of iterations at runtime. To simplify the problem, we increase or decrease the size of all the CF subtrees at the same time in Algorithm 6. We will have the evaluation of the auto-tuning strategy in Sect. 4.10.3.

Algorithm 6 Algorithm for searching the optimal way to divide a task graph into CF subtrees

```
Require: α1, ..., αm (CF root tasks in the initial partition)
Require: T (Processing time of an iteration under the initial partitioning)
Ensure: Optimal CF root tasks
1: int Tn = 0, Tc = T ; // New & current processing time
2: int EvalLarger = 1; //Eval. larger subtrees?
3: void EvaluateNewPartitioning () {
4:   Execute an iteration under the new partitioning ;
5:   Record the processing time Tn ;
6: }
7: SearchOptimalPartitioning () {
8: while CF root tasks have child tasks do
9:   Set child tasks of the current CF root tasks as the new CF root tasks ;
10:   EvaluateNewPartitioning() ;
11:   if Tn < Tc then
12:     Tc = Tn; Save new CF root tasks ;
13:     EvalLarger = 0 ; //Point 1, do not evaluate larger subtrees
14:   else
15:     break ;
16:   end if
17: end while
18: if EvalLarger == 1 then
19:   Restore CF root tasks to {α1, ..., αm} ; //Point 3, evaluate larger subtrees
20:   Tc = T ;
21:   while CF root tasks have parent tasks do
22:     Set parent tasks of the current CF root tasks as the new CF root tasks;
23:     EvaluateNewPartitioning() ;
24:     if Tn < Tc then
25:           Tc = Tn;
26:           Save new CF root tasks ;
27:     else
28:       break ;
29:     end if
30:   end while
31: end if
32: }
```

4.7 Triple-Level Work-Stealing Policy

In this section, we present how the scheduling policy in LAWS works so that the tasks are scheduled as required by the load-balanced task allocator and the cache-friendly task graph partitioner. Similar to the bi-tier work-stealing policy proposed to improve the shared cache utilization in Chap. 3, LAWS leverages a triple-level work-stealing policy to improve both the shared cache utilization and the NUMA memory system utilization.

Figure 4.9 gives the runtime architecture of LAWS on an M-socket architecture employed NUMA-based memory system, and illustrates the triple-level work-stealing policies in LAWS. In Fig. 4.9, the main memory is divided into M memory

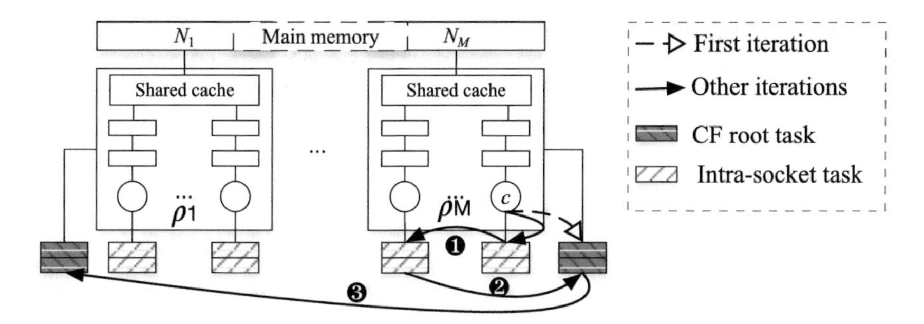

Fig. 4.9 Runtime architecture of LAWS on an M-socket architecture with NUMA-based memory system

nodes and node N_i is the local memory node of socket ρ_i. In each socket, core "0" is selected as the head core of the socket.

Observed from Fig. 4.9, for each socket, LAWS creates a *CF task pool* to store CF root tasks allocated to the socket and the tasks above the CF root tasks in the task graph. For each core, LAWS creates an *intra-socket task pool* to store the intra-socket tasks. Suppose a core c in socket ρ is free, in different phases, it obtains new tasks in different ways as follows.

In the first iteration of an iterative program (or the independent initialization phase if the program has the phase), there is no intra-socket task and all the tasks are pushed into CF task pools since the task graph has not been divided into CF subtrees. In the period, core c in the socket ρ can only obtain a new task from the CF task pool of ρ. Core c is not allowed to steal a task from other sockets because the data set of a task will be stored into the wrong memory node if it is stolen in the first iteration due to the first touch strategy.

Starting from the second iteration, the task graph of each iteration has been divided into multiple CF subtrees. Adopting triple level work-stealing policy, a free core c can steal a new task from three levels: intra-socket task pool of other cores in its socket ρ, CF task pool of ρ, and CF task pools of other sockets. More precisely, when c is free, it first tries to obtain a task from its own intra-socket task pool. If its own task pool is empty, **(1)** c tries to steal a task from the intra-socket task pools of other cores in ρ. If the task pools of all the cores in ρ are empty and c is the head core of ρ, **(2)** c tries to obtain a new CF root task from ρ's CF task pool. If ρ's CF task pool is also empty and c is the head core of ρ, **(3)** c tries to steal a task from CF task pools of other sockets. In other words, LAWS allows a socket to help other sockets execute their CF subtrees. Although ρ needs longer time to process the CF subtrees that are allocated to other sockets, the workload is balanced and the performance of memory-bound programs can be improved.

It is worth noting that the CF subtrees allocated to each socket are executed sequentially in LAWS. In other words, cores in the same socket are not allowed to execute tasks in multiple CF subtrees concurrently. This policy can avoid the situation that tasks in different CF subtrees pollute the shared caches with different data sets. A socket is only allowed to steal entire CF subtrees from other sockets for optimizing shared cache usage.

4.8 Theoretical Validation

In previous sections, we already present how LAWS splits the data set of a parallel program, and schedules the parallel tasks so that the data locality can be improved practically. Besides the actual design of LAWS, wn this section, we theoretically show that LAWS can improve the performance of memory-bound programs through locality-aware task scheduling compared with traditional random work-stealing.

Without loss of generality, in this section we analyze the divide-and-conquer programs that have tree-shaped task graphs. Divide-and-conquer programs are often the targeted programs of emerging work-stealing environments, e.g., TBB [27], Cilk++ [23] and X10 [22] etc. After analyzing memory-bound divide-and-conquer programs carefully, we find that they often have three main features. First, only leaf tasks physically access the data while other tasks divide the data set recursively into smaller pieces. Second, each leaf task only processes a small part of the whole data set of the program. Third, the execution time of a leaf task is decided by its data access time. Based on the three main features, the rest of this section proves that LAWS can improve the performance of memory-bound divide-and-conquer programs theoretically.

Consider a memory-bound program that runs on an M-socket architecture. Suppose a leaf task α in its task graph is responsible for processing data of S bytes and α still accesses B bytes of boundary data besides its own part of data. Let V_l and V_r represent the speeds (bytes/cycle) of a core to access data from local memory node and remote memory nodes respectively. Needless to say, $V_l > V_r$ in NUMA-based memory system. Table 4.2 lists the parameters used to validate that LAWS can improve the performance of memory-bound programs compared with the random work-stealing.

If we adopt a random work-stealing scheduler, e.g., MIT Cilk and TBB, to schedule the program, the probability that α can access all the data from local memory node is $1/M$. Therefore, the cycles expected for α to access all the needed data in traditional work-stealing (denoted by T_R) can be calculated by Eq. 4.7.

$$T_R = \frac{S+B}{V_l} \times \frac{1}{M} + \frac{S+B}{V_r} \times \frac{M-1}{M} \tag{4.7}$$

If we adopt LAWS to schedule the program, benefit from the load-balanced task allocator, α can access its own part of data from local memory node. As a consequence, the cycles needed by α to access all the needed data in LAWS (denoted by T_L) can be calculated by Eq. 4.8, because α also has a high chance to access its boundary data from local memory node.

Table 4.2 Parameters used in the theoretical validation

Parameters	Description
M	Number of sockets
B	Size of boundary data between two neighbor leaf tasks
S	Size of data allocated to a leaf task
V_l	Speed of a core to access data from the local memory node
V_r	Speed of a core to access data from a remote memory node
T_R	Time of reading data for a leaf task with random work-stealing
T_L	Time of reading data for a leaf task with LAWS

$$T_L \leq \frac{S}{V_l} + \frac{B}{V_r} \tag{4.8}$$

Deduced from Eqs. 4.7 and 4.8, we can get Eq. 4.9.

$$\begin{aligned} T_R - T_L &\geq \frac{(M-1) \times S - B}{M \times V_r} - \frac{(M-1) \times S - B}{M \times V_l} \\ &= \frac{(M-1) \times S - B}{M \times V_r} - \frac{(M-1) \times S - B}{M \times V_l} \\ &= (\frac{1}{M \times V_r} - \frac{1}{M \times V_l}) \times [(M-1) \times S - B] \end{aligned} \tag{4.9}$$

In Eq. 4.9, because $V_r < V_l$, we can know $\frac{1}{M \times V_r} - \frac{1}{M \times V_l} > 0$. Therefore, $T_R - T_L > 0$ if $(M-1) \times S - B > 0$ that is always true in almost all the divide-and-conquer programs empirically since a task's own data set (S) is always far larger than its boundary data (B). In summary, we prove that α needs shorter time to access all the needed data in LAWS compared with the random work-stealing scheduler.

Because leaf tasks need shorter time to access their data in LAWS than in traditional random work-stealing schedulers, LAWS can always improve the performance of memory-bound divide-and-conquer programs even when the optimization on reducing shared cache misses in LAWS is not taken into account.

4.9 Implementation Methodology

LAWS can be implemented in a similar way the CAB scheduler (Chap. 3) was implemented. As we presented in Sect. 3.5.2, two types of task-generating policies, child-first and parent-first, can be adopted when generating new tasks. In parent-first policy (called *help-first policy* in [17]), a core continually executes the parent task after spawning a new task. In child-first policy (called *work-first policy* in [5]), a core continually executes the spawned new task once the child is spawned. Parent-first policy works better when the steals are frequent, while child-first policy works better when the steals are infrequent [17].

During the first iteration, LAWS adopts the parent-first policy to generate new tasks, because it is difficult to collect the numbers of last level private cache misses caused by each task with the child-first policy. If a core is executing a task α, with the child-first policy, it is very likely the core will also execute some of α's child tasks before α is completed. In this case, the number of last level cache misses caused by α itself, which is used to calculate SOIDs of tasks, may not be collected correctly as it could include the number of last level private cache misses of α's child tasks. Starting from the second iteration, LAWS generates tasks above CF root tasks with the parent-first policy since the steals are frequent in the beginning of each iteration. LAWS generates intra-socket tasks with the child-first policy since the steals are infrequent in each CF subtree.

We have modified the *cilk2c* compiler to support both the parent-first and child-first task-generating policy while the original Cilk only support the child-first policy. If a task α is spawned in the first iteration, the task is spawned with the parent-first policy and is pushed to the appropriate CF task pool based on the method in Sect. 4.5. If α is spawned in the later iterations and it is an intra-socket task, LAWS spawns α with the child-first policy and pushes α into the intra-socket task pool of the current core. Otherwise, if α is a CF root task or a task above CF root tasks, and it is allocated to socket ρ, it is spawned with the parent-first policy and pushed into ρ's CF task pool.

4.10 Performance Evaluation of LAWS

In this section, we evaluate the performance of the introduced locality-aware work-stealing scheduler: LAWS. In the beginning, we introduce the two experimental hardware platforms used in the experiment. Then, on each experimental platform, we present the experimental results respectively. More precisely, we present the performance of LAWS for the memory-bound benchmarks, the effectiveness of the load-balanced task allocator and the cache-friendly task graph partitioner respectively, the scalability and the overhead of LAWS.

4.10.1 Experimental Platforms

We use both a Intel-based sever and an AMD-based server to evaluate the performance of LAWS. Table 4.3 lists the detailed hardware configurations. In the

Table 4.3 Configurations of the experimental platforms

AMD-based server	CPU	AMD Opteron 8380
	Num of sockets	4
	Cores per socket	4
	L2 cache (per core)	512 KB
	L3 cache (per socket)	6 MB
	DRAM	16 GB
	Operating System	Linux 3.2.0-14
Intel-based server	CPU	Intel Xeon X7560
	Num of sockets	4
	Cores per socket	8 cores (16 HW threads)
	L2 cache (per core)	2 MB
	L3 cache (per socket)	24 MB
	DRAM	64 GB
	Operating system	Linux 3.13.0-13

Intel-based server, Intel hyper-threading (HT) technology that delivers two processing threads per physical core is disabled.

We compare the performance of LAWS with the performance of Cilk [5] and CAB-GTO that is presented in Chap. 3. Cilk uses the pure child-first policy to spawn and schedule tasks. Similar to LAWS, CAB-GTO also packs the task graph of a parallel program into subtrees to reduce shared cache misses in MSMC architectures. Once a task graph is packed in CAB-GTO, the partitioning cannot be adjusted at runtime even the partitioning is not optimal. In addition, CAB-GTO did not consider the underlying NUMA memory system.

For fairness in comparison, we also implement CAB-GTO by modifying Cilk and we have improved CAB-GTO so that it also allocates the data evenly to all the memory nodes in the first iteration as LAWS does. The Cilk programs run with CAB-GTO and LAWS without any modification. In our experiment, the number of workers (i.e., threads) launched in Cilk, CAB-GTO and LAWS equals to the number of physical cores in the hardware platform. Furthermore, to avoid any performance variation due to OS level thread scheduling, we pin each worker with an individual hardware core.

To evaluate the performance of LAWS in different scenarios, we use benchmarks listed in Table 4.4 that have both regular task graph and irregular task graph in the experiment. Since there are no standardized large-scale benchmarks available for work-stealing schedulers so far, most of the benchmarks are examples in the MIT Cilk package. We port the other benchmarks in the same way the examples of MIT Cilk are developed. The benchmarks we used are the same as in previous papers [5, 18]. According to our experiment in Sect. 4.10.4, for current benchmarks, the larger the makespan, the better LAWS performs, which indicates the potential benefit of LAWS for large-scale benchmarks.

Table 4.4 Benchmarks used in the experiments

Name	Bound type	Description
Heat/Heat-ir	Memory-bound	2D heat distribution algorithm
SOR/SOR-ir	Memory-bound	Successive over-relaxation algorithm
GE/GE-ir	Memory-bound	Gaussian elimination algorithm
9P/9P-ir	Memory-bound	2D 9-point stencil computing algorithm
6P/6P-ir	Memory-bound	3D 6-point stencil computing algorithm
25P/25P-ir	Memory-bound	3D 25-point stencil computing algorithm
Mandelbrot	CPU-bound	Algorithm of calculating mandelbrot set
Queens(15)	CPU-bound	N-queens problem
FFT	CPU-bound	Fast fourier transform algorithm
GA	CPU-bound	Island model of genetic Algorithm
Knapsack	CPU-bound	0–1 knapsack problem

Heat-ir, *GE-ir*, *SOR-ir*, *9P-ir*, *6P-ir* and *25P-ir* implement the same algorithm as their counterparts respectively, except their task graphs are irregular. We create the programs with irregular task graphs in the same way in Sect. 3.8.2. If all the nodes (except the leaf tasks) in the DAG have the same branching degrees, the task graph is regular. All benchmarks are compiled with "-O2". For each test, every benchmark is run ten times and the average execution time is reported as the result.

4.10.2 Performance of LAWS

Figure 4.10 shows the performance of memory-intensive benchmarks in Cilk, CAB-GTO and LAWS on an AMD-based server and an Intel-based server.

On the AMD-based server, for *Heat*, *Heat-ir*, *SOR*, *SOR-ir*, *9P* and *9P-ir* the input data used is a 8096 × 1024 matrix. For *GE* and *GE-ir*, the input data used is a 2048 × 2048 matrix due to algorithm constraint. For *6P*, *6P-ir*, *25P* and *25P-ir*, the input data is a 8096 × 64 × 64 3D matrix. On the Intel-based server, for *Heat*, *Heat-ir*, *SOR*, *SOR-ir*, *9P* and *9P-ir* the input data used is a 8096 × 4096 matrix. For *GE* and *GE-ir*, the input data used is a 8192 × 8192 matrix due to algorithm constraint. For *6P*, *6P-ir*, *25P* and *25P-ir*, the input data is a 8096 × 128 × 128 3D matrix.

As we can see from Fig. 4.10, on the AMD-based server, LAWS can significantly improve the performance of benchmarks compared with Cilk while the performance improvement ranges from 23.5 to 54.2%. CAB-GTO can also improve the performance of benchmarks up to 19.6% compared with Cilk. On the Intel-based server, LAWS can also significantly improve the performance of benchmarks compared with Cilk while the performance improvement ranges from 12.5 to 48.6%. CAB-GTO can also improve the performance of benchmarks up to 28.1% compared with Cilk.

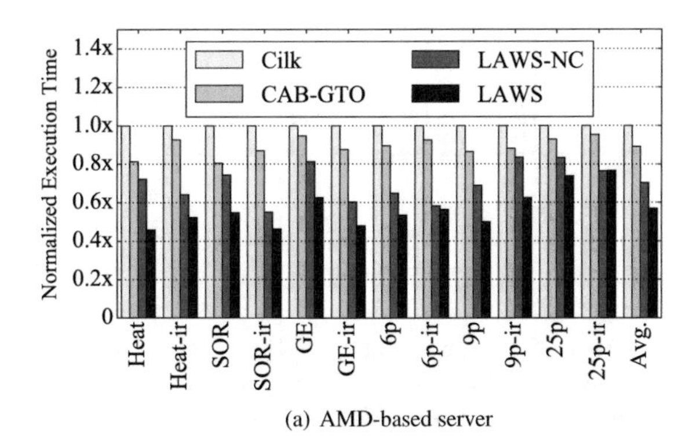

(a) AMD-based server

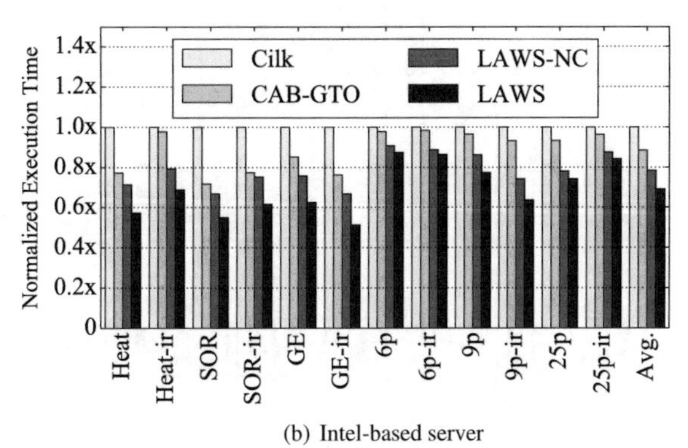

(b) Intel-based server

Fig. 4.10 The performance of memory-bound benchmarks in Cilk, CAB-GTO, LAWS-NC and LAWS

In MSMC architectures, the performance of a memory-bound application is decided by the *straggler socket* that seldom access data from its local memory node because the tasks in the straggler socket need the longest time to access their data. During the execution of a memory-bound application, any socket in the MSMC architecture can be the straggler socket.

To explain why LAWS outperforms both Cilk and CAB-GTO for memory-bound applications on both AMD-based server and Intel-based server, we also collect the shared cache misses (Event "LLC_MISSES") and the local memory accesses of the straggler socket using "libpfm" library in Linux. For each benchmark, Table 4.5 lists its shared cache misses and the local memory accesses of the straggler socket in Cilk, CAB-GTO and LAWS.

Observed from Table 4.5, we can find that the shared cache (L3) misses are reduced and the local memory accesses of the straggler socket are prominently increased in LAWS compared with Cilk and CAB-GTO. Since LAWS schedules tasks to the sockets where the local memory nodes store their data, the tasks can access their data from local memory node and thus the local memory accesses have been significantly increased. Furthermore, since LAWS packs tasks allocated to each socket into CF subtrees to preserve shared data in shared cache, the shared cache misses are also reduced.

Table 4.5 Shared cache misses and local memory accesses of the straggler socket

Regular benches				Heat	SOR	GE	6P	9P	25P
	AMD-based server	L3 cache misses	Cilk	5.72E8	1.15E9	2.20E8	2.52E9	5.73E8	2.48E9
			CAB-GTO	5.31E8	1.07E9	1.47E8	2.42E9	5.39E8	2.38E9
			LAWS	4.62E8	1.01E9	2.91E7	2.38E9	5.05E8	2.34E9
		Local memory accesses	Cilk	1.61E7	3.28E7	6.1E6	8.15E7	1.72E7	8.32E7
			CAB-GTO	2.13E7	4.14E7	4.5E6	1.01E8	2.19E7	9.06E7
			LAWS	2.58E7	5.71E7	6.5E5	1.519E8	2.72E7	1.25E8
	Intel-based server	L3 Cache misses	Cilk	1.19E9	2.39E9	7.82E8	3.48E9	9.41E8	2.31E9
			CAB-GTO	1.1E9	2.17E9	7.68E8	3.11E9	9.27E9	2.24E9
			LAWS	9.96E8	2.01E9	4.96E8	3.07E9	9.23E8	2.22E9
		Local memory accesses	Cilk	9062	79239	112344	110218	7576	269769
			CAB-GTO	17105	72522	133341	106469	4885	201611
			LAWS	27563	99510	145126	131165	27643	373682
Irregular benches				Heat-ir	SOR-ir	GE-ir	6P-ir	9P-ir	25P-ir
	AMD-based Server	L3 cache misses	Cilk	5.74E8	1.01E9	2.30E8	2.54E9	5.77E8	2.48E9
			CAB-GTO	5.42E8	8.86E8	1.13E8	2.36E9	4.69E8	2.37E9
			LAWS	5.05E8	8.76E8	2.87E7	2.34E9	4.46E8	2.35E9
		Local memory accesses	Cilk	1.72E7	2.9E7	5.64E6	7.44E7	1.53E7	8.15E7
			CAB-GTO	1.86E7	3.04E7	3.58E6	9.73E7	1.93E7	8.58E7
			LAWS	2.75E7	3.93E7	4.7E5	1.347E8	2.48E7	1.18E8
	Intel-based Server	L3 cache misses	Cilk	1.17E9	2.43E9	8.35E8	2.85E9	9.42E8	2.35E9
			CAB-GTO	1.05E9	2.19E9	8.16E8	2.78E9	9.41E9	2.27E9
			LAWS	9.9E8	1.97E9	4.97E8	2.38E9	9.33E8	2.23E9
		Local memory accesses	Cilk	8444	71723	121324	86822	4142	266848
			CAB-GTO	10309	79019	141726	92804	5250	248204
			LAWS	24569	102833	147161	125722	19406	381267

Only for *GE* and *GE-ir* on AMD-based server, the local memory accesses of the straggler socket are not increased in LAWS. This is because their input data is small enough to be put into the shared cache directly. In this situation, most tasks can access the data from the shared cache directly and do not need to access the main memory any more. Because the L3 cache misses are prominently reduced, LAWS can still significantly improve the performance of *GE* and *GE-ir* compared to Cilk and CAB-GTO. The performance improvement of the benchmarks in CAB-GTO is due to the reduced shared cache misses. However, since CAB-GTO cannot divide a task graph optimally like LAWS, it still has more shared cache misses than LAWS as shown in Table 4.5.

Careful readers may find that CAB-GTO performs much worse here than in Chap. 3. While CAB-GTO can only improve the performance of benchmarks up to 19.6% here, it can improve their performance up to 74.4% in Sect. 3.8.2. The reduction of performance improvement of CAB-GTO comes from the much larger input data set used in this chapter. This result matches with the findings in Chap. 3. That is, with the increasing of the size of the input data set, the percentage of shared data among tasks decreases and the effectiveness of CAB-GTO degrades in consequence.

4.10.3 Effectiveness of Cache-Friendly Task Graph Partitioner

To evaluate the effectiveness of the cache-friendly task graph partitioner in LAWS, we compare the performance of LAWS with LAWS-NC, a scheduler that only schedules each task to the socket where the memory node stores its part of data but does not further pack the tasks into CF subtrees.

From Fig. 4.10 we find that LAWS-NC performs better than Cilk and CAB-GTO. This is because most tasks in LAWS-NC can access their data from local memory nodes. However, since tasks are not packed into CF subtrees for optimizing shared cache in LAWS-NC, LAWS-NC incurs more shared cache misses and performs worse than LAWS.

To evaluate the auto-tuning approach (Algorithm 6) proposed to optimally pack tasks into CF subtrees, Fig. 4.11 gives the execution time of 200 iterations of all the benchmarks with irregular task graphs in LAWS on AMD-based server and the execution time of 200 iterations of all the benchmarks with regular task graphs in LAWS on Intel-based server. All the other benchmarks give similar result. From the figure we find that the execution time of an iteration in all the benchmarks is significantly reduced after the optimal partitioning is found in several iterations.

In summary, the cache-friendly task graph partitioner in LAWS is effective and the auto-tuning algorithm for searching the optimal partitioning of tasks in Algorithm 6 works also fine.

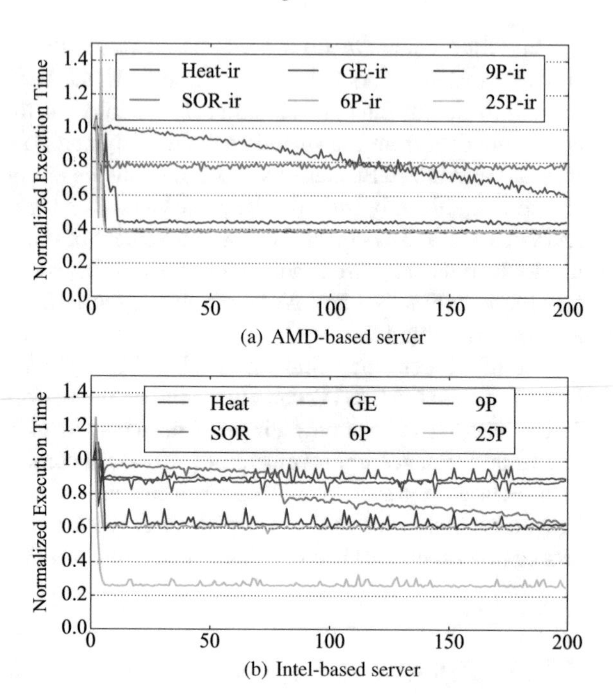

Fig. 4.11 Execution time of each iteration in all the benchmarks in LAWS on AMD-based server and Intel-based server

4.10.4 *Scalability of LAWS*

To evaluate scalability of LAWS, we compare the performance of benchmarks with different input data sizes in Cilk, CAB-GTO and LAWS.

During the execution of all the benchmarks, every task divides its data set into several parts by rows to generate child tasks unless the task meets the cutoff point (i.e., the rows of a leaf task, and 8 rows is used in the experiment). Since the data set size of the leaf tasks affects the measurement of scalability, we ensure that the data set size of the leaf tasks is constant by using a constant cutoff point for the leaf tasks. On AMD-based server, if the input data is an $x \times y$ 2D matrix, we set $y = 1024$ for all the input 2D matrix. If the input data is an $x \times y \times z$ 3D matrix, we set $y = 64$ and $z = 64$ for all the input 3D matrix. On Intel-based server, if the input data is an $x \times y$ 2D matrix, we set $y = 4096$ for all the input 2D matrix. If the input data is an $x \times y \times z$ 3D matrix, we set $y = 128$ and $z = 128$ for all the input 3D matrix.

We only adjust the x of the input matrices in the experiment. In this way, we can measure the scalability of LAWS without the impact of the granularity of the leaf tasks. In all the following figures, the x-axis represents the x of the input matrixes.

We use *Heat-ir* and *6P* on AMD-based server, *SOR-ir* and *9P* on Intel-based server as benchmarks to evaluate the scalability of CAB-GTO in scenario that applications with a regular task graph and an irregular task graph. All the other benchmarks have similar results. We omit them here due to the limited space.

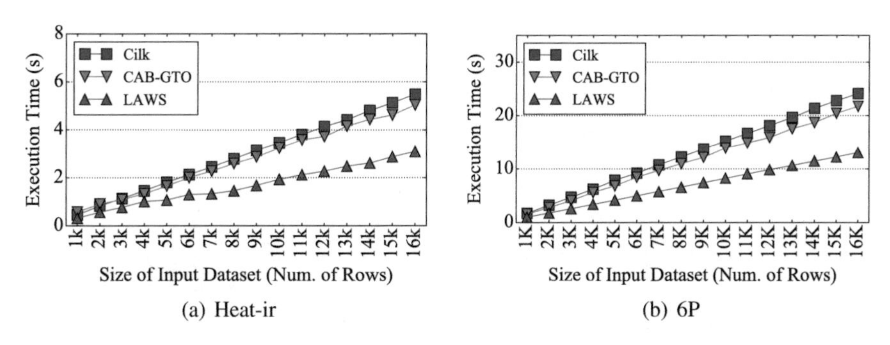

Fig. 4.12 Performance of Heat-ir and 6P with different input data sizes on the AMD-based server

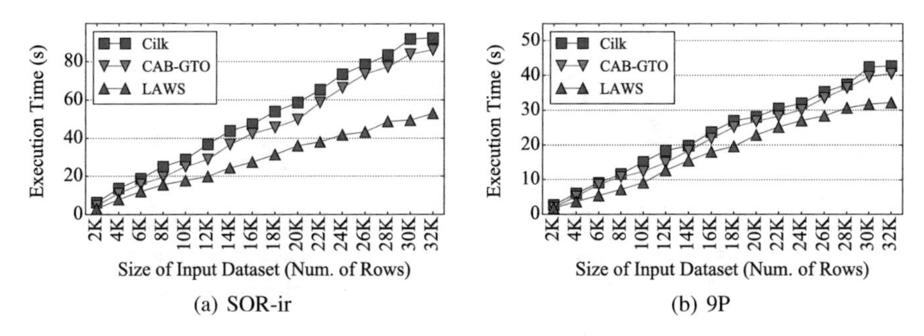

Fig. 4.13 Performance of SOR-ir and 9P with different input data sizes on the Intel-based server

Figures 4.12 and 4.13 show the performance of benchmarks with different input data sizes in Cilk, CAB-GTO and LAWS. We can find that *Heat-ir*, *6P*, *SOR-ir* and *9P* achieve the best performance in LAWS for all input data sizes. When the input data size is small (i.e., $x = 1k$), LAWS reduces 30.4% execution time of *Heat-ir* and reduces 36.6% execution time of *6P* compared with Cilk on AMD-based server. When the input data size is large (i.e., $x = 16k$), LAWS reduces 43.6% execution time of *Heat-ir* and reduces 45.8% execution time of *6P* compared with Cilk on AMD-based server. When the input data size is small (i.e., $x = 2k$), LAWS reduces 52.5% execution time of *SOR-ir* and reduces 34.9% execution time of *9P* compared with Cilk on Intel-based server. When the input data size is large (i.e., $x = 32k$), LAWS reduces 42.7% execution time of *SOR-ir* and reduces 24.7% execution time of *9P* compared with Cilk on Intel-based server.

In Figs. 4.12 and 4.13, the execution time of benchmarks in Cilk, CAB-GTO and LAWS increases linearly with the increasing of their input data sizes. Since their execution time increases much slower in LAWS than in Cilk and CAB-GTO, for all the input data sizes, LAWS can always reduce the execution time of memory-bound applications. In summary, LAWS is scalable in scheduling both regular task graphs and irregular task graphs.

Corresponding to Figs. 4.12, 4.13, 4.14 and 4.15 show the L3 cache misses and the local memory accesses of the straggler socket in executing *Heat-ir*, *6P*, *SOR-ir* and *9P*

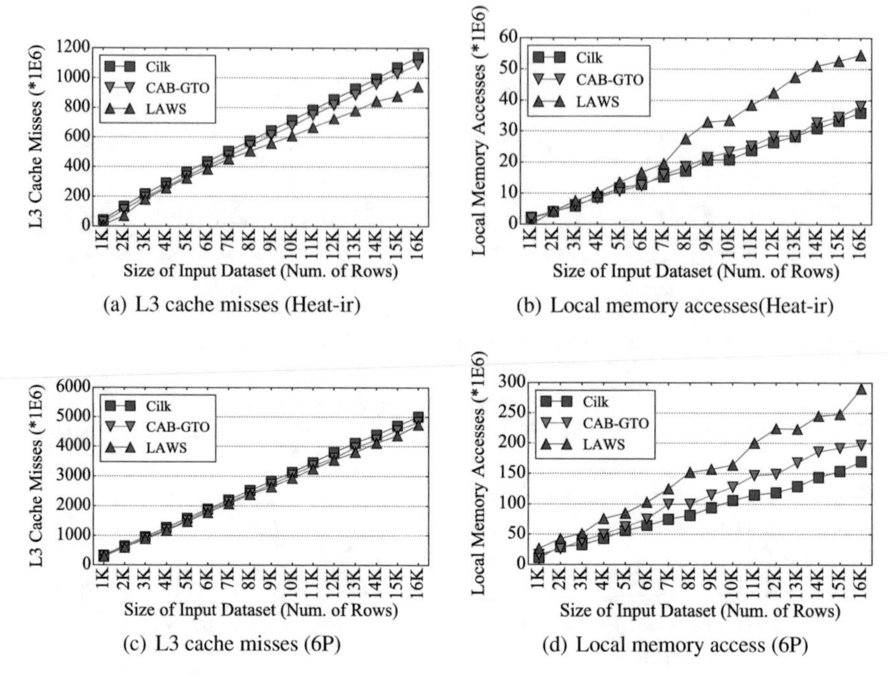

Fig. 4.14 L3 cache misses and local memory accesses of the straggler socket in Heat-ir and 6P on AMD-based server

with different input data sizes. Observed from the figure, we can find that the shared cache misses are reduced, while the local memory accesses of the straggler socket are increased in LAWS. When the input data size is small (i.e., $x = 1k$), LAWS can reduce 82% L3 cache misses and increase 132.1% local memory accesses compared with Cilk. When the input data size is large (i.e., $x = 16k$), LAWS can reduce 17.3% L3 cache misses and increase 70.6% local memory accesses compared with Cilk.

Figures 4.14 and 4.15 further explain why LAWS performs much better than CAB-GTO. Since LAWS can optimally pack tasks into CF subtrees through auto-tuning, it can reduce more L3 cache misses of memory-bound benchmarks than CAB-GTO. In addition, since LAWS can schedule a task to the socket where the local memory node stores its data, it significantly increases local memory accesses. The two key advantages of LAWS result in the better performance of LAWS.

Careful readers may observe from Fig. 4.15 that LAWS failed to reduce the last level shared cache misses for *9P* on Intel-based server. However, because LAWS significantly improve the local memory access, *9P* still performs much better than Cilk and CAB-GTO.

As we all know, if the input data of a memory-bound program is small, the shared cache is big enough to store the input data. In this case, if the shared cache misses are greatly reduced, the performance of memory-bound programs can be improved. If the input data is large, the performance bottleneck of the program is the time of

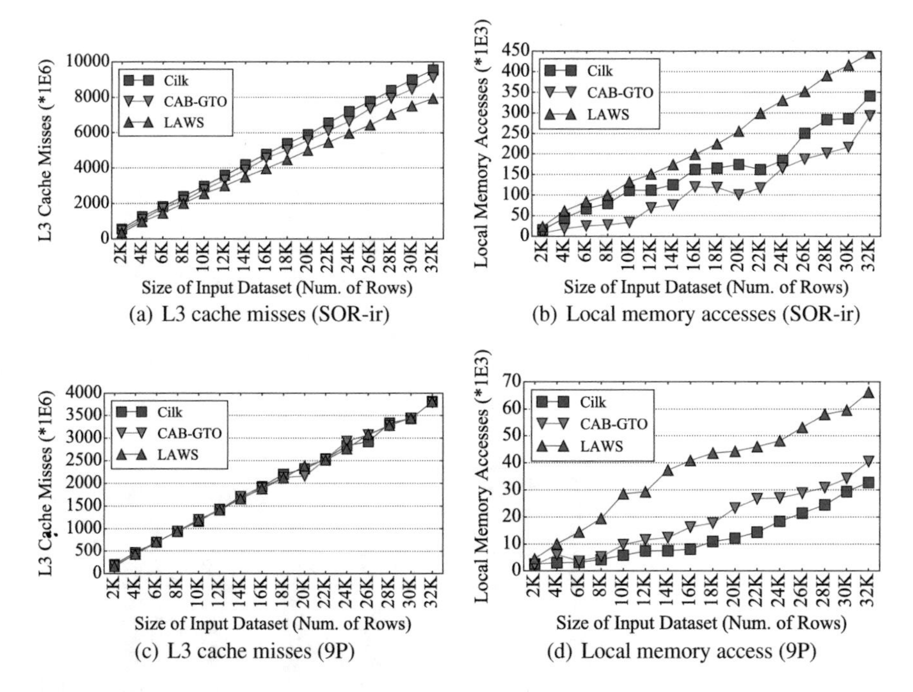

Fig. 4.15 L3 cache misses and local memory accesses of the straggler socket in SOR-ir and 9P on Intel-based server

reading data from main memory. Therefore, CAB-GTO performs efficient when the input data size is small but performs poor when the input data size is large. On the contrary, because LAWS can increase more local memory accesses when input data size gets larger, it performs even better when the input data is large. This feature of LAWS is promising as the data size of a problem is becoming larger and larger.

4.10.5 Overhead of LAWS

Because LAWS aims to increase local memory accesses and reduce shared cache misses, LAWS is neutral for CPU-bound programs. Based on the runtime information, if LAWS finds that a program is CPU-bound, LAWS schedules tasks of the program in traditional work-stealing. In Chap. 5, we will introduce the technique that can be used to improve the performance of CPU-bound programs by balancing workloads among cores.

Figure 4.16 shows the performance of several CPU-bound applications in Cilk, CAB-GTO and LAWS on AMD-based server and Intel-based server. The applications in this experiment are examples in Cilk package. By comparing the performance

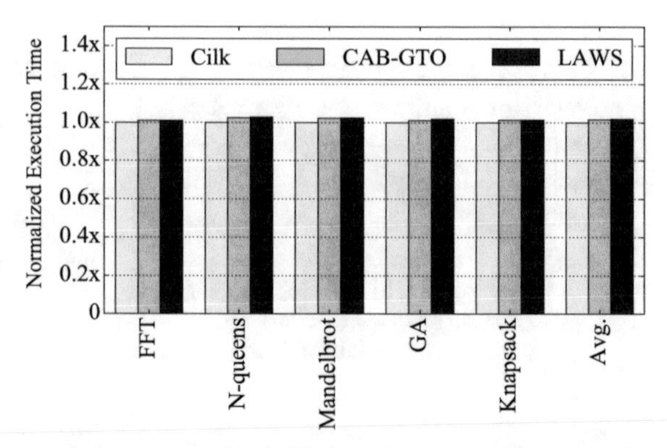

(a) AMD-based server

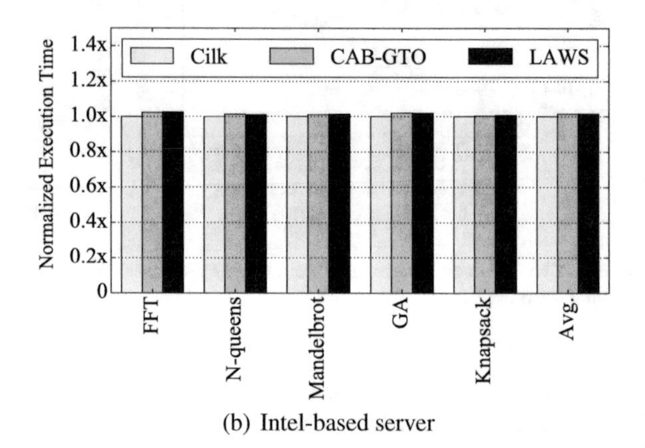

(b) Intel-based server

Fig. 4.16 Performance of CPU-bound benchmarks in Cilk, CAB-GTO and LAWS on AMD-based server and Intel-based server

of CPU-bound applications in Cilk, CAB-GTO and LAWS, we can find the extra overhead of LAWS.

Observed from Fig. 4.16, we see the extra overhead of LAWS is negligible (less than 3% of the overall execution time) compared with Cilk and CAB-GTO. The extra overhead of LAWS mainly comes from the overhead of distributing data to all the memory node evenly and the profiling overhead in the first iteration of a parallel program, when LAWS can determine if the program is CPU-bound or memory-bound based on the profiling information.

4.10.6 Applicability of LAWS

LAWS assumes that the task graphs of different iterations in an iterative program are the same. The assumption is true for most programs. Even if a program does not satisfy this assumption, LAWS can still ensure that every task can access its data from local memory node since the load-balanced task allocator allocates tasks to sockets in each iteration independently according to their data set in the current iteration. However, in this situation, the optimization on shared cache utilization is not applicable since the optimal partitioning for the past iterations may not be optimal for future iterations due to the change of the task graph. In summary, even the above assumption is not satisfied, LAWS can improve the performance of memory-bound programs due to the increased local memory accesses.

As LAWS is neutral for CPU-bound programs, LAWS decides at runtime if an application is CPU-bound based on profiled information. When LAWS collects cache misses in the first iteration, it also collects the number of retired instructions of the task through performance monitoring counter. If the *cache miss intensity* (i.e., cache misses per instruction) of a task is smaller than a given threshold, the task is labelled as CPU-bound. If most tasks of an application are CPU-bound, the application is regarded as CPU-bound by LAWS.

4.11 Summary

Traditional work-stealing schedulers suffer from the shared cache polution and the small number of local memory accesses in MSMC architectures with NUMA-based memory system. To solve the two problems, we have proposed the LAWS scheduler, which consists of a load-balanced task allocator, a cache-friendly task graph partitioner and a triple-level work-stealing scheduler. The load-balanced task allocator evenly distributes the data set of a program to all the memory nodes and allocates a task to the socket where the local memory node stores its data for increasing local memory accesses. Based on auto-tuning, for each socket, the cache-friendly task graph partitioner can optimally pack the allocated tasks into CF subtrees to optimize shared cache usage. The triple-level work-stealing scheduler schedules tasks in the same CF subtree among cores in the same socket and makes sure that each socket executes its CF subtrees sequentially. Experimental results show that LAWS can improve the performance of memory-bound programs up to 54.2% on AMD-based experimental platform and up to 48.6% on Intel-based experimental platform compared with traditional work-stealing schedulers. Furthermore, the extra overhead of LAWS for CPU-intensive applications is negligible.

4.11.1 Chapter Highlights

The following highlights of this chapter could be of your interest:

- We systematically analyze the poor data locality problem in emerging MSMC architecture employed NUMA-based memory system.

- We propose a load-balanced task allocation policy that automatically allocates a task to the particular socket where the local memory node stores its data and that can balance the workload among sockets.
- We propose a cache-friendly task graph partitioner that can further pack a task graph into *Cache Friendly Subtrees* (CF subtrees) for optimizing shared cache usage based on online-collected information and auto-tuning.
- We propose a triple-level work-stealing policy to schedule tasks accordingly so that a task can access its data from either the shared cache or the local memory node other than the remote memory nodes.
- We demonstrate that LAWS significantly improves the performance of memory-intensive applications. The experiment shows that LAWS can achieve a performance gain of up to 54.2% on AMD-based experimental platform and up to 48.6% on Intel-based experimental platform for memory-intensive applications.

References

1. U. Acar, G. Blelloch, and R. Blumofe. The data locality of work stealing. Theory of Computing Systems, 35(3):321–347, 2002.
2. AMD. *BIOS and Kernel Developer Guide (BKDG) For AMD Family 10 h Processors*. AMD (2010).
3. E. Ayguadé, N. Copty, A. Duran, J. Hoeflinger, Y. Lin, F. Massaioli, X. Teruel, P. Unnikrishnan, and G. Zhang. The design of OpenMP tasks. IEEE TPDS, 20(3):404–418, 2009.
4. R. D. Blumofe. *Executing Multithreaded Programs Efficiently*. Ph.D. thesis, MIT, September 1995.
5. R. D. Blumofe, C. F. Joerg, B. C. Kuszmaul, C. E. Leiserson, K. H. Randall, and Y. Zhou. Cilk: An efficient multithreaded runtime system. Journal of Parallel and Distributed Computing, 37(1):55–69, 1996.
6. M. Castro, L. G. Fernandes, C. Pousa, J.-F. Méhaut, and M. S. de Aguiar. NUMA-ICTM: A parallel version of ICTM exploiting memory placement strategies for NUMA machines. In *IPDPS*, pp. 1–8, (2009).
7. Q. Chen and M. Guo. Adaptive workload aware task scheduling for single-ISA multi-core architectures. *ACM Transactions on Architecture and Code Optimization*, 11(1) (2014).
8. Q. Chen, Y. Chen, Z. Huang, and M. Guo. WATS: Workload-aware task scheduling in asymmetric multi-core architectures. In *IPDPS*, pp. 249–260 (2012).
9. Q. Chen, M. Guo, and Z. Huang. CATS: Cache aware task-stealing based on online profiling in multi-socket multi-core architectures. In *ICS*, pp. 163–172 (2012).
10. Q. Chen, Z. Huang, M. Guo, and J. Zhou. CAB: Cache-aware bi-tier task-stealing in multi-socket multi-core architecture. In *ICPP*, pp. 722–7320 (2011).
11. Q. Chen, and M. Guo. Locality-aware work stealing based on online profiling and auto-tuning for multisocket multicore architectures. ACM Transactions on Architecture and Code Optimization, 12(2):22, 2015.
12. R. Cole and V. Ramachandran. Analysis of randomized work stealing with false sharing. In *IPDPS*, pp. 985–989 (2013).
13. M. Frigo, C. E. Leiserson, and K. H. Randall. The implementation of the Cilk-5 multithreaded language. In *PLDI*, pp. 212–223 (1998).
14. T. Gautier, J. V. Lima, N. Maillard, and B. Raffin. XKaapi: A runtime system for data-flow task programming on heterogeneous architectures. In *IPDPS*, pp. 1299–1308 (2013).
15. T. Gautier, J. V. F. Lima, N. Maillard, B. Raffin, et al. Locality-aware work stealing on Multi-CPU and Multi-GPU architectures. In *MULTIPROG* (2013).

16. A. Gerasoulis and T. Yang. A comparison of clustering heuristics for scheduling directed acyclic graphs on multiprocessors. Journal of Parallel and Distributed Computing, 16(4):276–291, 1992.
17. Y. Guo, R. Barik, R. Raman, and V. Sarkar. Work- first and help-first scheduling policies for async-finish task parallelism. In *IPDPS*, pp. 1–12 (2009).
18. Y. Guo, J. Zhao, V. Cave, and V. Sarkar. SLAW: a scalable locality-aware adaptive work–stealing scheduler. In *IPDPS*, pp. 1–12 (2010).
19. L. V. Kale and S. Krishnan. *CHARM++: a portable concurrent object oriented system based on C++*. ACM (1993).
20. G. Karypis and V. Kumar. A fast and high quality multilevel scheme for partitioning irregular graphs. SIAM Journal on Scientific Computing, 20(1):359–392, 1998.
21. T. Kielmann, R. F. Hofman, H. E. Bal, A. Plaat, and R. A. Bhoedjang. Magpie: Mpis collective communication operations for clustered wide area systems. In *Proceeding 7th ACM SIGPLAN Symposium on Principles and Practice of Parallel Programming, Atlanta, GA*. Citeseer (1999).
22. J. Lee and J. Palsberg. Featherweight X10: a core calculus for async-finish parallelism. In *PPoPP*, pp. 25–36 (2010).
23. C. Leiserson. The Cilk++ concurrency platform. In *DAC*, pp. 522–527 (2009).
24. A. Muddukrishna, P. A. Jonsson, V. Vlassov, and M. Brorsson. Locality-aware task scheduling and data distribution on numa systems. In *OpenMP in the Era of Low Power Devices and Accelerators*, pp. 156–170. Springer (2013).
25. L. L. Pilla, C. P. Ribeiro, D. Cordeiro, A. Bhatele, P. O. Navaux, J.-F. Méhaut, L. V. Kalé, et al. Improving parallel system performance with a NUMA-aware load balancer. *TR-JLPC-11-02* (2011).
26. J.-N. Quintin and F. Wagner. Hierarchical work-stealing. In *EuroPar*, pp. 217–229 (2010).
27. J. Reinders. *Intel threading building blocks*. Intel (2007).
28. M. Shaheen and R. Strzodka. NUMA aware iterative stencil computations on many-core systems. In *IPDPS*, pp. 461–473 (2012).
29. S. Sridharan, G. Gupta, and G. S. Sohi. Holistic run-time parallelism management for time and energy efficiency. In *ICS*, pp. 337–348 (2013).
30. B. Vikranth, R. Wankar, and C. R. Rao. Topology aware task stealing for on-chip NUMA multi-core processors. Procedia Computer Science, 18:379–388, 2013.
31. R. Yang, J. Antony, A. Rendell, D. Robson, and P. Strazdins. Profiling directed NUMA optimization on Linux systems: A case study of the Gaussian computational chemistry code. In *Proceedings of the International Parallel and Distributed Processing Symposium*, pp. 1046–1057, Anchorage, Alaska, USA. IEEE (2011).
32. R. M. Yoo, C. J. Hughes, C. Kim, Y.-K. Chen, and C. Kozyrakis. Locality-aware task management for unstructured parallelism: a quantitative limit study. In *SPAA*, pp. 315–325 (2013).
33. R. Van Nieuwpoort, T. Kielmann, and H. E. Bal. Efficient load balancing for wide-area divide-and-conquer applications. In *ACM SIGPLAN Symposium on Principles and Practice of Parallel Programming*. Citeseer (2001).

Chapter 5
Dynamic Load Balancing for Asymmetric Multi-core Architecture

Abstract In Chaps. 3 and 4, we have introduced the techniques to improve the performance of memory-bound applications on multi-socket architecture. In this chapter, on the other hand, we introduce the scheduling techniques proposed to improve the performance of CPU-bound applications. On Symmetric Multi-Core (SMC) architecture in which all cores provide equal performance, traditional random work-stealing performs well. However, while single-ISA Asymmetric Multi-Core (AMC) architectures have shown high performance as well as power efficiency, current parallel programming environments do not perform well on AMC because they are designed for SMC architectures. The random task scheduling policies used in current parallel programming environments, such as work-sharing and work-stealing, can result in unbalanced workloads in AMC and severely degrade the performance of parallel applications. Essentially, it is a NP-hard problem to find the optimal task scheduling on an AMC architecture. In order to balance the workloads of parallel applications in AMC, in this chapter, we introduce an Asymmetric-Aware Task Scheduling (AATS) methodology.

5.1 Chapter Organization

We have introduced the advantages of Asymmetric Multi-Core (AMC) architecture in Chap. 1. However, despite the rapid development of the AMC technology, current parallel programming environments as listed below, still assume all cores provide equal performance. Due to this assumption, parallel applications cannot utilize the asymmetric cores of an AMC architecture effectively. Most current parallel programming environments adopt either work-sharing or work-stealing policies for task scheduling. By dynamically scheduling the parallel tasks, the workloads can be balanced in multi-core architectures. However, both work-stealing and work-sharing (refer to Chap. 2) do not consider tasks' workloads when allocating tasks to differ-

Part of contents in this chapter has been published through ACM Transactions on Architecture and Code Optimization. Reprinted from Ref. [6], with permission from ACM. Figures 5.1, 5.7, 5.8, 5.9 and 5.10 in this chapter have been published through ACM Transactions on Architecture and Code Optimization. Reprinted from Ref. [6], with permission from ACM

© Springer Nature Singapore Pte Ltd. 2017

Q. Chen and M. Guo, *Task Scheduling for Multi-core and Parallel Architectures*,
https://doi.org/10.1007/978-981-10-6238-4_5

ent cores, which is not a problem for symmetric cores but can cause unbalanced workloads among asymmetric cores. For example, a long task may be scheduled to a slow core, while a short task is executed by a fast core. This problem of unbalanced workloads can significantly degrade the performance of parallel applications.

To this end, in this chapter, we introduce the techniques proposed to improve the performance of parallel programs (especially, batch-based programs and pipeline-based programs) on single-ISA AMC architectures where different types of cores in an AMC have the same Instruction Set Architecture (ISA). All the tasks of a parallel program can be executed by any core in a single-ISA AMC architecture directly.

This chapter is organized as follows. Section 5.2 describes the problem of unbalanced workloads in AMC and the proposed solutions. Section 5.3 discusses related work. Section 5.6 presents an Asymmetric-Aware Task Scheduling (AATS) methodology. Section 5.7 introduces the history-based task allocation in AATS. Section 5.8 presents the preference-based work-stealing in AATS. Section 5.9 gives the implementation details. Section 5.10 evaluates AATS, provides experimental results, performance evaluation, and limitations of AATS. Section 5.11 summarizes the contributions of AATS.

5.2 Problem Formulation

In this section, we use an example to explain and formulate the problem of the unbalanced workloads in AMC with the traditional asymmetric-unaware task scheduling policies. Suppose a parallel application has four independent parallel tasks: γ_1, γ_2, γ_3 and γ_4. Assume that the application runs on an AMC architecture with one fast core (c_0) and three slow cores (c_1, c_2 and c_3) as shown in Fig. 5.1. Suppose γ_1, γ_2, γ_3 and γ_4 take times f_1, f_2, f_3 and f_4 on the fast core c_0 respectively and $f_1 > f_2 > f_3 > f_4$. As the counterpart, γ_1, γ_2, γ_3 and γ_4 take times s_1, s_2, s_3 and s_4 on the slow cores respectively. We can reasonably deduce that $s_1 > f_1, s_2 > f_2, s_3 > f_3$ and $s_4 > f_4$. For easing of description and discussion, we further assume $f_1 > s_2, f_1 > s_3$ and $f_1 > s_4$.

Figure 5.1 shows two possible allocations of the independent parallel tasks γ_1, γ_2, γ_3 and γ_4 to the asymmetric cores. Figure 5.1a is an optimal allocation where γ_1 is allocated to the fast core c_0 and the shorter tasks are allocated to the slow

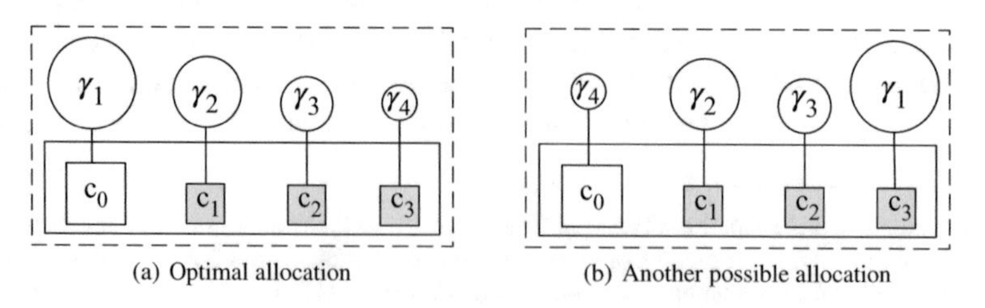

Fig. 5.1 Comparison of two possible allocations of the independent tasks γ_1, γ_2, γ_3 and γ_4. Allocating long tasks (γ_1 and γ_2) to slow cores and vice versa may result in the poor performance of a parallel program

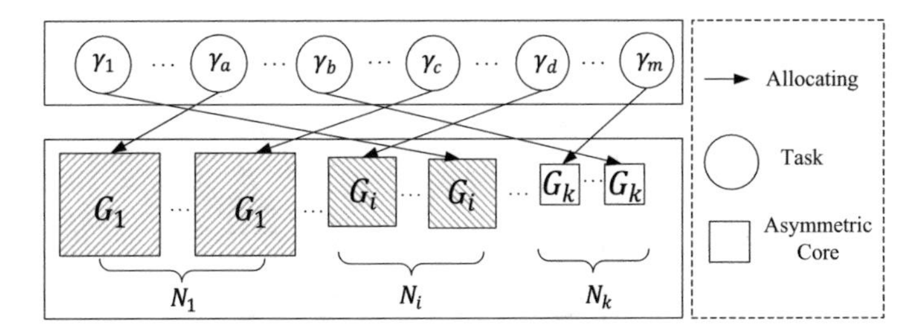

Fig. 5.2 The optimal task allocation problem in AMC. Allocate m independent tasks with different workloads to k c-groups with different computational capacities

cores. The makespan (i.e., the overall completion time) for γ_1, γ_2, γ_3 and γ_4 is $T_{opt} = max\{f_1, s_2, s_3, s_4\} = f_1$. Because $f_1 < s_1$, we deduce that $T_{opt} < s_1$.

However, without considering the performance asymmetric feature of AMC and the character of different tasks, traditional task scheduling policies proposed for symmetric multi-core architecture such as work-stealing, γ_1, γ_2, γ_3 and γ_4 are likely to be allocated as in Fig. 5.1b. In the allocation, γ_3 is allocated to the fast core but the long task γ_1 is scheduled to a slow core. In this case, the makespan for γ_1, γ_2, γ_3 and γ_4 is $T_{trad} = max\{s_1, s_2, f_3, s_4\} \geq s_1 > f_1 = T_{opt}$. Obviously, allocating a long task to a slow core would often degrade the overall performance seriously in traditional task scheduling policies.

Without loss of generality, Fig. 5.2 illustrates the general problem of optimal task allocation in AMC. Suppose there are m independent tasks (γ_1, ..., γ_m) with different workloads and an AMC with k types of cores. We group cores of the same type into a core group (denoted as c-group). We use G_1, ..., G_k to represent the k c-group in descending order of their computational capacities, and use N_i $(1 \leq i \leq k)$ to represent the number of cores in G_i. The problem can be expressed as *how to divide the m tasks into k groups that are assigned to the k c-groups respectively, so that the makespan is minimum?* Once tasks are assigned to c-groups, many existing task scheduling policies (e.g., work-sharing and work-stealing) can be adopted to balance workloads among symmetric cores in the same c-group.

5.3 Existing Solutions

5.3.1 Task Snatching Technique

A straightforward way to improve the random scheduling on AMC is allowing idle fast cores to snatch tasks from slow cores [1]. For example, with this rescuing policy, for the situation in Fig. 5.1b, c_0 is allowed to snatch γ_1 from c_3 after finishing γ_4. Figure 5.3 shows the way of scheduling tasks with the task snatching technique.

As shown in Fig. 5.3, suppose c_0 snatches γ_1 from c_3 after finishing γ_4 (which takes time f_4). Core c_0 still needs $(\frac{s_1 - f_4}{s_1}) \times f_1$ to finish γ_1 because c_3 has only finished

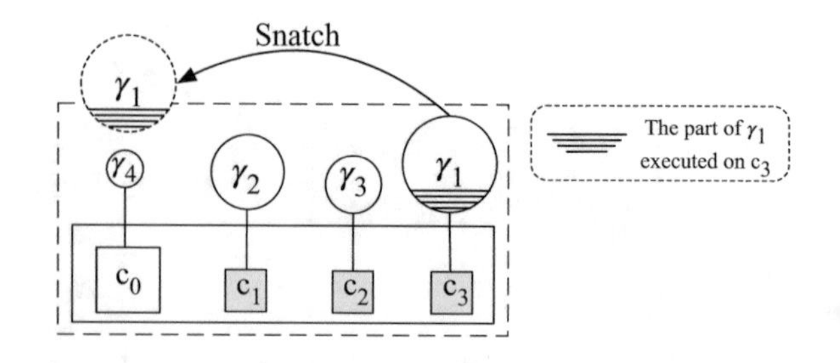

Fig. 5.3 Scheduling tasks with the task snatching technique

$\frac{s_1-f_4}{s_1}$ of γ_1. Let Δ_s represent the time of the snatching operation. Then the overall time for c_0 to finish both γ_4 and γ_1 is $f_4 + \frac{s_1-f_4}{s_1} \times f_1 + \Delta_s$. Therefore, with the task snatching technique, the makespan for γ_1, γ_2, γ_3 and γ_4 is $T_{res} = max\{f_4 + \frac{s_1-f_4}{s_1} \times f_1 + \Delta_s, s_2, s_3, f_4\}$.

Because $f_4 + \frac{s_1-f_4}{s_1} \times f_1 + \Delta_s - f_1 > \frac{f_1}{s_1} \times f_4 + \frac{s_1-f_4}{s_1} \times f_1 - f_1 + \Delta_s = \frac{f_4}{s_1} \times f_1 + \frac{s_1-f_4}{s_1} \times f_1 - f_1 + \Delta_s = \Delta_s$, we can deduce that $f_4 + \frac{s_1-f_4}{s_1} \times f_1 + \Delta_s > f_1$. In addition, since f_1 is larger than s_2, s_3 and f_4, $T_{res} = f_4 + \frac{s_1-f_4}{s_1} \times f_1 + \Delta_s > f_1 = T_{opt}$ and the rescuing policy is still not as efficient as the optimal allocation.

Furthermore, since $T_{res} - T_{bad} = f_4 + \frac{s_1-f_4}{s_1} \times f_1 + \Delta_s - s_1 = (s_1 - f_4) \times (\frac{f_1}{s_1} - 1) + \Delta_s$ and $(s_1 - f_4) \times (\frac{f_1}{s_1} - 1) < 0$, if the system knows the execution time of each task on all the cores and Δ_s is not too large, the snatching policy can improve the performance of random scheduling. However, the execution time of the tasks on different cores are unknown to the existing task schedulers. Therefore, idle fast cores have to snatch tasks randomly and thus the snatching policy will still suffer from the randomness in the random scheduling. For example, in Fig. 5.1b, with the random snatching, the worst case could be that c_0 first snatches γ_2 and γ_3 before snatching γ_1, where the makespan is larger.

In summary, the rescuing policy, task snatching, performs worse than the optimal scheduling that directly allocates each task to the appropriate core. To achieve the optimal scheduling, the knowledge of tasks' execution time on different cores is essential to optimal task scheduling in AMC. This knowledge can help a scheduler allocate long tasks to fast cores, which is often optimal. It can also help idle fast cores to steal or snatch the long tasks if steal and snatch are necessary. By comparing the task snatching policy with the optimal scheduling, we can conclude that *an initial optimal allocation based on the knowledge of workloads is more crucial to the makespan than the snatching policy that tries to rescue a non-optimal allocation.* To this end, we introduce *Asymmetric-Aware Task Scheduling* (AATS) that can find the initial near-optimal task allocation.

5.3.2 CAMP

Besides the task snatching technique, many studies have been done to explore optimal task scheduling in different parallel platforms [5, 23]. Especially, in AMC, many studies on scheduling focus on resource allocation at the OS level [2, 7, 14, 20]. They aim to achieve high system throughput by balancing the hardware resources (e.g., cores and caches) among different programs. For instance, Saez et al. [22] proposed a OS-level thread scheduler, *CAMP*, to optimize system throughput by devoting fast cores to run high-speedup applications in AMC.

5.3.2.1 Utility Factor

In CAMP, a new metric *Utility Factor* (UF), which produces a single value that approximates how much an application will improve its performance if its threads are allowed to occupy all the fast cores available on an AMC. The metric is designed to help the scheduler picks the best threads to run on fast cores in non-trivial cases. For instance, suppose a workload of a CPU-intensive application A with two runnable threads and a less CPU-intensive workload B with a single thread are running on an AMC with one fast core. In this case, it is not immediately clear which thread is the best candidate for running on the fast core. On the one hand, dedicating the fast core to a thread of A may bring smaller performance improvements to A than dedicating the fast core to B, because a smaller part of A will be running on fast cores in the former case. On the other hand, A is more CPU-intensive, so running it on the fast core may be more profitable than dedicating the fast core to the less CPU-intensive application B. By comparing utility factors across threads the scheduler should be able to identify the most profitable candidates for running on fast cores.

The UF of an application on an AMC architecture is calculated as follows in CAMP. Given an AMC with N_{fc} fast cores, Eq. 5.1 calculates the utility factor UF of a parallel application *app*. In the equation, N_t is the number of threads in the application *app*, which is visible to the operating system. SF_{app} is the average speedup factor of *app*'s threads when running on a fast core relative to a slow core

$$UF = \frac{SF_{app}}{(max(1, N_t - (N_{fc} - 1)))^2} \tag{5.1}$$

In order to understand this equation, we first to consider the case where *app* has only a single thread. In this case, UF=SF; in other words the utility factor is equal to the speedup that *app* will experience from running on a fast core relative to a slow core. If *app* is multithreaded and all its threads were running on fast cores, it would achieve the speedup of SF. In that case, the denominator is equal to one and UF=SF. However, if the number of threads is greater than the number of fast cores, then only some of the threads will run on fast cores and the overall utility factor will be less than SF. In order to account for that, in the equation, SF is divided by one greater than the number of threads that would not be running on fast cores: $N_t - (N_{fc} - 1)$. Finally, a quadratic factor is introduced in the denominator, because the experiment shows that if some of the threads experience the speedup because of running on fast cores and others do not, the overall application speedup is smaller than the portion of speedup achieved by threads running on fast cores. That is because threads running on fast cores must synchronize with the threads running on slow cores, so they do not fully contribute to the application-wide speedup. Introducing the quadratic factor in the formula enables to account for that effect rather accurately. Saez et al. [22] has proved the effectiveness of this equation experimentally.

5.3.2.2 Scheduling Algorithm

After the utility factors of all the applications are calculated, CAMP decides which threads to place on cores of different types based on their individual utility factors. In order to achieve this purpose, threads are categorized into three classes: LOW, MEDIUM, and HIGH according to their utility factors. Threads falling in the HIGH utility class will be allocated to fast cores.

Specifically, if the number of high-utility threads is larger than the number of fast cores, the fast cores are shared among these threads equally, using a round-robin mechanism. Otherwise, if the number of high-utility threads is smaller than the number of fast cores, there are some idle fast cores remaining and they will be used for running medium-utility threads or low-utility threads. In contrast to the threads in the HIGH utility class, fast cores will not be shared equally for threads in the MEDIUM and LOW utility classes. Sharing the cores equally implies cross-core migrations as threads are moved between fast and slow cores. These migrations hurt performance, especially for memory-intensive threads, because threads may lose their last-level cache state as a result of migrations.

In addition, for a parallel application, its threads that execute a sequential phase will be designated to a special class SEQUENTIAL_BOOSTED. These threads will get the highest priority for running on fast cores: this provides more opportunities to accelerate sequential phases. Only high-utility threads, however, will be assigned to the SEQUENTIAL_BOOSTED class. Medium- and low-utility threads will belong to their regular class despite running sequential phases. Because these threads do not use fast cores efficiently, it is not worthwhile to give them an elevated status. Threads placed in the SEQUENTIAL_BOOSTED class will remain there for the duration of *amp_boost_ticks*, a configurable parameter. After that, they will be downgraded to their regular class, as determined by the utility factor, to prevent them from monopolizing the fast core. The class-based scheme followed by CAMP relies on two utility thresholds, *lower* and *upper*, which determine the boundaries between the LOW, MEDIUM and HIGH utility classes. The lower threshold is used to separate the LOW and MEDIUM classes, the upper threshold is used to separate the MEDIUM and HIGH classes.

As we discussed above, CAMP is a thread-level scheduler. Because tasks in the same task-based programs can often achieve similar speedup ratios on fast cores,

CAMP is not applicable to improve the performance of a single parallel program. Therefore, CAMP did not considered the scheduling problem in parallel applications that this chapter will address in AMC.

5.3.3 Bias Scheduling

Koufaty et al. [12] proposed a bias scheduling which matches threads to the right type of cores through dynamically monitoring the bias of the threads in order to maximize the system throughput. In this work, each application is given a *bias*, which reflects the core type that best suits its resource needs. By dynamically monitoring application bias, the operating system is able to match threads to the core type that can maximize system throughput. Bias scheduling takes advantage of this by influencing the existing scheduler to select the core type that bests suits the application when performing load balancing operations.

5.3.3.1 Bias

Koufaty et al. [12] defines application bias as the type of core that the operating system would prefer to run threads of the application at a particular time. More specifically:

- A thread has a small core bias if its speedup from running on a big core compared to a small core would be modest.
- A thread has a big core bias if its speedup from running on a big core compared to a small core would be large.

The definition of what constitutes a modest or large speedup is dependent on the characteristics of the cores. In order to select values for these speedups, an operating system might compare the performance of both cores on a battery of tests and select a specific range in the curve to map a large speedup (e.g. the upper quartile).

It is worth noting that application bias is not static. While an application might have a certain bias overall, it can change as the application goes through different phases of the computation. Different threads from the same application might also have different bias.

In order to identify the application bias, we first classify core stall cycles broadly into *internal stalls* and *external stalls*. Internal stalls are caused by the lack of resources internal to the core (an execution unit, a load port), competition on those resources (a TLB or private cache miss) or natural inaccuracies on them (branch mis-prediction). Most of these events are short in duration and can often be hidden by out-of-order microarchitectures. However, they are numerous and include many sources, leading to pipeline stalls even in the most CPU-bound tests. External stalls External stalls are caused by access to resources external to the core. They include shared last level caches, memory and I/O. These events are significantly less frequent than internal stalls. However, their latency can be orders of magnitude larger than internal stalls.

External and internal stalls can be used as a strong predictor of the application bias. Generally speaking, as proved by Koufaty et al. [12], applications with Cycles-Per-Instruction (CPI) dominated by either internal or external stalls have a small core bias. If neither internal nor external stalls dominate the CPI, the application is dominated by execution cycles. Given the nature of the two core types, the big core is very likely to outperform the small core significantly, therefore we consider this type of application to have a big core bias.

In order to compute application bias, performance monitoring hardware available in modern processors can be used to measure internal and external stalls. Application bias changes as the amount of stalls fall below or climb above predefined thresholds, switching an application between small or big core bias.

5.3.3.2 Bias Scheduling Algorithm

Bias scheduling algorithm is built on top of the existing scheduler of operating system. It does not change the existing scheduler of operating system in ways that would dictate when to run a thread or change any system properties that the scheduler is trying to maintain such as fairness, responsiveness or real time constraints. It works by preferably scheduling applications on the core type that best suits its bias.

In order to implement bias scheduling, the operating system is modified to support fast-core first scheduling and asymmetry-aware load balancing. The fast-core first scheduling enables maximum performance by scheduling first to idle big cores. The asymmetry-aware load balancing schedules work in big cores proportional to an estimated ratio of big/small performance.

It is worth noting that bias scheduling can be performed on top of any existing scheduling algorithm. Its design tries to minimize changes to the existing scheduler by focusing on two areas: *imbalanced system* and *balanced system* as described below.

- On an imbalanced asymmetric system, the scheduler tries to migrate a thread from the busiest core to the idlest core. Bias scheduling does not change the way in which these cores are selected. Once they have been identified, the required load that has the highest bias is migrated to the destination core. If the cores have the same type, bias is irrelevant. In some instances the scheduler cannot find a thread with an appropriate bias, and defaults to migrating any thread just as it does without bias.
- On a balanced asymmetric system, the load balance is periodically checked and nothing is done if the system is balanced. In such case, the load balancer/scheduler inspects the load on the run-queue of the critical big cores looking for a thread that has a small core bias. If such thread is found, it then searches the small cores looking for a thread that has a big core bias. It then performs a thread swap on

their run-queues. This process is started only in big cores, but it works similarly if started on a small core.

Finally, the bias scheduling does not modify the initial thread allocation. While it is possible to allow for user annotations or historic tables for initial scheduling, that is complementary to the bias scheduling. In practice, however, Koufaty et al. [12] have found that threads show bias fairly quickly, so it might not be necessary to improve initial allocation. For instance, threads can be allocated to fast/slow cores during the initial allocation according to their bias.

Furthermore, bias scheduling only hooks into the existing scheduler during load balancing. It is also limited to influence the core where a thread will be run based on bias and has no effect whatsoever on when the thread is run, making it transparent to other parts of the scheduler that select what to run based on priorities, fairness and other scheduler design considerations. Therefore, implementation on top of any existing general purpose scheduler design is simplified significantly.

5.3.4 Age-Based Scheduling

Lakshminarayana et al. [13] proposed age-based scheduling to improve the performance of multi-threaded applications on AMC architecture. Age-based scheduling targets multi-threaded applications that follow a simple fork-join model as shown in Fig. 5.4, in which each dotted line is a thread.

In the model, the main thread of the application forks several threads that perform the actual computation. These threads may encounter multiple barriers during their whole lifetime. After all the threads complete their work, they join to the main thread. The goal of the age-based scheduling is to schedule the threads such that they all reach their next mile-stones, which could be either a barrier or termination, at the same time. In this case, the performance of the application could be optimal. It tries to do this by utilizing all the available cores, while exploiting the fast cores to accelerate threads that are lagging behind. Because of the asymmetric nature of the cores, threads that execute on slow cores automatically lag behind threads executing on fast cores. Many well-designed applications, such as *Blackscholes* and *Swaptions* in Parsec benchmark suite, fall in this category naturally.

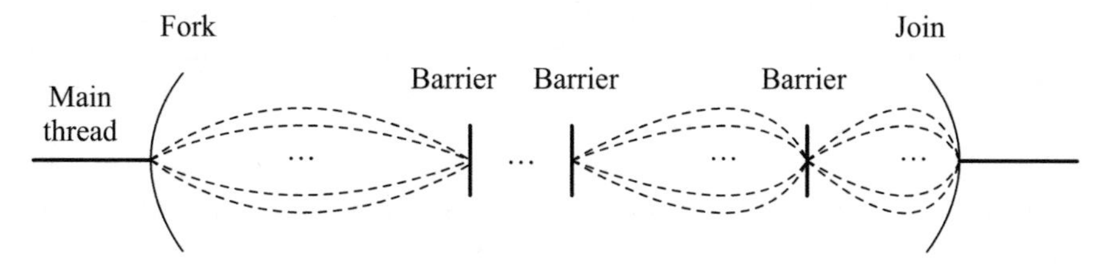

Fig. 5.4 Application model assumed by the age-based scheduling

5.3.4.1 LJFCF Scheduling Policy

Based on the age of the threads, Lakshminarayana et al. [13] proposed the *Age Based Longest Job Fast Core First* (LJFCF) policy. The policy predicts how long a thread is from the next barrier (termination if there is no more barrier). Based on the distance, the threads that have the longest predicted distance to the next milestone is assigned to the fast cores. The challenging problem here is to exactly predict the remaining time of a thread to the next milestone.

Fortunately, LJFCF policy does not need to know the absolute values of distances of different threads, but only need to know the relative values of distances of threads. In LJFCF policy, the prediction of relative values of distances to the next milestone is based on the insight that threads that are created together usually have the same lengths. Thus, threads that are created together are predicted to have the same distances to their next milestones at the time of their creation and threads that are created later are predicted to have longer distances to their next milestones. When a thread reaches a milestone, its distance to the next milestone is predicted and the newly predicted distance is used for thread to core assignment.

LJFCF is applied whenever one of the following occurs: (1) a thread is created, (2) a core goes idle because all the threads assigned to the core have either terminated or blocked, or (3) periodic timer for reassignment expires. If the above cases occur, it means that the whole system enters a new state, thread reassignment is necessary to achieve the best performance of an multi-threaded application.

Meanwhile, in order to apply LJFCF policy in the operating system, the periodic load balancing in traditional operating system should be replaced with periodic reassignment. While load balancing is typically done for each core at different points in time, reassignment is done at the same time for all cores.

5.3.4.2 Age Prediction

Lakshminarayana et al. [13] have propose two practical ways: *prediction*, and *profiling*, to calculate the remaining execution time, denoted by *rem_exe*, of a thread.

If LJFCF policy uses prediction to determine the distance of each thread from its next milestone, it predicts that the distance between any two successive milestones (creation, barriers and termination) of any thread is the same and the common distance value is predicted to be very large. This implies that after crossing a milestone all threads are predicted to have the same distance to their next milestone. As a thread executes, the predicted remaining distance to the next milestone reduces according to its progress. When it is invoked, the predicted remaining distance to the next milestone is used as *rem_exe* to assign threads to cores.

If the method of profiling is adopted by LJFCF policy, the distance between any two successive milestones are not the same across different threads. By profiling the application with sample inputs, the average distance between milestones is determined for each thread. These average distances are fed to the scheduler as a set of ratios before the application is run. Based on these ratios, the c. Threads with lower ratios are predicted to have shorter distances between milestones.

Based on the above techniques, Algorithm 7 shows the detailed algorithm of the LJFCF policy.

Algorithm 7 The LJFCF Policy in Age-based Scheduling

Require: : The *rem_exe* of all the threads;
Require: : N_t (number of threads), N_c (number of cores) ;
1: sort the threads in the decreasing order of their *rem_exe* ;
2: **if** $N_t \leq N_c$ **then**
3: Allocate each thread to a core while threads with larger *rem_exe* are allocated to fast cores ;
4: **end if**
5: **if** $N_t > N_c$ **then**
6: $avg\_rem\_exe\_core_i = (\sum_{b=0}^{N_t-1} rem\_exe_b / \sum_{b=0}^{N_c-1} core\_perf_b) \times core\_perf_i$; //Compute the average *rem_exe* that must be assigned to each core i ;
7:
8: **for all** thread $thread_j$ in sorted order **do**
9: $rem\_rem\_exe_i = avg\_rem\_exe\_core_i$ - $rem\_exe\_core_i$;
10: $k = max(rem\_rem\_exe_i)$; //identify the core $core_k$ for which the difference between the *rem_exe* to be assigned and the *rem_exe* actually assigned is the highest ;
11: assign $thread_j$ to $core_k$;
12: **if** number of threads yet to be assigned is less than or equal to the number of cores without any assignments **then**
13: make 1:1 assignment of threads to cores ;
14: break ;
15: **end if**
16: **end for**
17: **end if**

In the algorithm, rem_exe_b is the remaining distance of thread $thread_b$ to the next milestone; $core\_perf_b$ is the normalized performance of core $core_b$; $avg\_rem\_exe\_core_i$ is the average *rem_exe* that should be assigned to core $core_i$ for the best performance; $rem\_exe\_core_i$ is the remaining distance of all the threads allocated to $core_i$. Observed from the algorithm, we can find that the age-based scheduling policy tries to balance the remaining distance of all the active threads to the cores so that they can complete at the same time. In this way, the performance of multi-threaded applications can be improved.

5.3.5 Speed-Based Balancing

Hofmeyr et al. [10] proposed a speed balancing algorithm to manage the migration of threads so that each thread has a fair chance to run on the fastest core available. Instead of balancing the workloads, the algorithm balances the time of a thread executing on faster and slower cores.

5.3.5.1 Emerging Load Balancer in Linux

In emerging Linux, the *Load* on each core is defined as the number of tasks in the per-core run-queue. Linux attempts to balance the load (queue lengths) system wide by periodically invoking the load balancing algorithm on every core to pull tasks from longer to shorter queues, if possible. Linux is topology-aware by scheduling tasks in term of *scheduling domains*. The scheduling domains form a hierarchy that reflects the way hardware resources are shared: SMT hardware context, cache, socket and NUMA domain. Balancing is done progressing up the hierarchy and at each level, the load balancer determines how many tasks need to be moved between two groups to balance the sum of the loads in those groups. If the balance cannot be improved (e.g. one group has 3 tasks and the other 2 tasks) Linux will not migrate any tasks.

Emerging load balancer may fail to balance run-queues due to the constraints on migrating tasks. In particular, the balancer will never migrate the currently running task, and it will resist migrating "cache hot" tasks, where a task is designated as cache-hot if it has executed recently (around 5ms) on the core (except for migration between hardware contexts on SMT). This is a simple locality heuristic that ignores actual memory usage. If repeated attempts (typically between one and two) to balance tasks across domains fail, the load balancer will migrate cache-hot tasks. If even migrating cache-hot tasks fails to balance groups, the balancer will wake up the kernel-level *migration thread*, which walks the domains from the base of the busiest core up to the highest level, searching for an idle core to push tasks to.

5.3.5.2 Advantages of Speed-Based Balancing

Suppose all the threads of a parallel application have the same workload and they have to synchronize their execution. In this case, the performance of the application is that of the slowest thread and variation in "execution speed" of any thread negatively affects the overall system utilization and performance. A particular thread will run slower than others due to running on the core with the longest queue length, sharing a core with other threads with higher priority or running on a core with lower computational power (slower speed).

Let us use an example to explain the benefit of speed balancing. Consider a parallel application with N threads running on M homogeneous cores, $N > M$. Assume that threads will execute for the same amount of time S seconds and balancing executes every B seconds. Intuitively, S captures the duration between two program barrier or synchronization points. With Linux load balancing, the total program running time under these assumptions is at most $(T + 1) \times S$, the execution time on the slow cores.

With fair per-core schedulers, the average thread speed is $f \times \frac{1}{T} + (1 - f) \times \frac{1}{T+1}$, where f represents the fraction of time the thread has spent on a fast core. In default Linux, since the queue-length based load balancing will not migrate threads so the overall application speed is that of the slowest thread $\frac{1}{T+1}$. Ideally, each thread should spend an equal fraction of time on the fast cores and on the slow cores. In this case, the asymptotic average thread speed becomes $\frac{1}{2\times T} + \frac{1}{2\times(T+1)}$ which amounts to a possible speedup of $\frac{1}{2\times T}$.

According to the above analysis, it is much faster than the default queue-length based scheduler in Linux. By letting all the threads to run for the same time on the fast core, the performance of the multi-threaded applications can be improved.

5.3.5.3 Algorithm of Speed-Based Balancing

In order to actually implement speed balancing, Hofmeyr et al. [10] define $speed = \frac{t_{exec}}{t_{real}}$, in which t_{exec} is the elapsed execution time and t_{real} is the wall clock time. This measure directly captures the share of CPU time received by a thread and can be easily adapted to capture behavior in asymmetric systems. It is simpler than using the inverse of queue length as a speed indicator because that requires weighting threads by priorities, which can have different effects on running time depending on the task mix and the associated scheduling classes. Using the execution time based definition of speed is a better measure than the length of run queue.

During the implementation, Hofmeyr et al. [10] uses a helper thread (*balancer*) running on each core (named the *local* core). Each balancer operates independently without any global synchronization. Periodically, a *balancer* will wake up, check for imbalances, correct them by pulling threads for a slower core to the *local* core (if possible) and then goes to sleep again. The period over which the balancer sleeps (*balance interval*) determines the frequency of migrations. In more detail, each *balancer* performs the following steps.

1. For every thread t_i, on local core c_j, it computes the speed s_j^i over the elapsed balance interval.
2. It computes the local core speed s_j over the balance interval as the average of the speeds of all the threads on the local core: $s_j = average(s_j^i)$.
3. It computes the global core speed s_{global} as the average speed over all cores: $s_{global} = average(s_j)$.
4. It attempts to balance if the local core speed is greater than the global core speed: $s_j > s_{global}$.

The only interaction between the *balancers* is the mutual exclusion on the variable s_{global}. The *balancer* attempts to balance the time of each thread on fast core by searching for a suitable remote core c_k to pull threads from. A remote core c_k is suitable if its speed is less than the global speed ($s_k < s_{global}$) and it has not recently been involved in a migration. This heuristic has the side effect that it allows cache hot threads to run repeatedly on the same core. Once it finds a suitable core c_k, the balancer pulls a thread from the remote core c_k to the local core c_j. The balancer chooses to pull the thread that has migrated the least in order to avoid creating tasks that migrate repeatedly.

Lacking of global synchronization between *balancer*, the current implementation of speed-based balancing is not able to guarantee that each migration will be optimal. In another word, each *balancer* makes its own decision independently, which may result in a migration from the slowest core to a core that is faster than average, but not actually the fastest core.

5.3.6 Scheduling on AMC with Hardware Support

Besides the above discussed pure software techniques, some other researchers proposed techniques to improve the performance of parallel applications on AMC with extra hardware support. These techniques cannot be directly used in emerging real system computers. For instance, Suleman et al. [24], proposed ACS (Accelerated Critical Sections) to accelerate the execution of critical sections by migrating the threads with critical sections to fast cores. Similar to ACS, Joao et al. [11] proposed a cooperative software-hardware mechanism, BIS (*Bottleneck Identification and Scheduling*) that identify and accelerate the most critical bottlenecks. BIS identifies the most critical bottlenecks by measuring the number of cycles threads have to wait for each bottleneck and accelerates the bottlenecks using fast cores on an AMC architecture. In addition, while BIS needed to add some structures in hardware, in this chapter, we introduce a pure software approach.

As observed from the schedulers discussed above (CAMP, bias scheduling, age-based scheduling, and speed-based balancing), they improve the performance of applications by scheduling threads. In their assumption, the work of a thread is fixed once it is created, which is not true in emerging task-based applications scheduled with work-sharing or work-stealing. They are not applicable for emerging applications, where different tasks have different workloads.

5.4 Theoretical Ideal Task Scheduling

Without loss of generality, we generalize the task allocation problem, assuming the execution time of tasks on all the cores are known. In other words, we assume that a *Duration Table* (DT) for the parallel program that has m tasks on the AMC that has k c-groups has already known in Table 5.1. In the table, the item t_{ji} in row γ_j and column G_i is the expected execution time of γ_j on a core in c-group G_i.

Based on the duration table in Table 5.1, the following theorem provides theoretical guidance to optimal task allocation.

Theorem 5.1 *For tasks $\gamma_1, ..., \gamma_m$, if $\gamma_{p_{i-1}+1}, ..., \gamma_{p_i}$ $(1 \le i \le k, p_0 = 0, p_k = m)$ are allocated to c-group G_i, their makespan is minimum only when $p_1, ..., p_{k-1}$ satisfy*

$$\sum_{n=1}^{p_1} t_{n1} : ... : \sum_{n=p_{i-1}+1}^{p_i} t_{ni} : ... : \sum_{n=p_{k-1}+1}^{m} t_{nk} = N_1 : ... : N_i : ... : N_k \tag{5.2}$$

Table 5.1 Duration Table (DT) of the program that has m tasks on an AMC architecture with k c-groups

Tasks	C-groups					
	G_1	G_2	...	G_i	...	G_k
γ_1	t_{11}	t_{12}	...	t_{1i}	...	t_{1k}
γ_2	t_{21}	t_{22}	...	t_{2i}	...	t_{2k}
...	...	...	...	...	...	...
γ_j	t_{j1}	t_{j2}	...	t_{ji}	...	t_{jk}
...	...	...	...	...	...	...
γ_m	t_{m1}	t_{m2}	...	t_{mi}	...	t_{mk}

Moreover, the task allocation is optimal and the optimal makespan $T_{opt} = \frac{\sum_{n=1}^{p_1} t_{n1}}{N_1} = \dots = \frac{\sum_{n=p_{i-1}+1}^{p_i} t_{ni}}{N_i} \dots = \frac{\sum_{n=p_{k-1}+1}^{m} t_{nk}}{N_k}$.

Proof Straightforward. If tasks are divided into groups in Eq. 5.2, the workloads are balanced among the k c-groups in terms of the computation capacities of the cores in different c-groups. Since all the workloads are fully balanced during the time period T_{opt} and the lower bound is achieved, this task allocation is optimal. Therefore, the execution time for the group of tasks allocated on the k c-groups can be calculated as $\frac{\sum_{n=1}^{p_1} t_{n1}}{N_1} = \dots = \frac{\sum_{n=p_{i-1}+1}^{p_i} t_{ni}}{N_i} \dots = \frac{\sum_{n=p_{k-1}+1}^{m} t_{nk}}{N_k} = T_{opt}$.

However, it is not feasible to find the ideal solution to Theorem 5.1 because they may not exist in real situations. Even if they exist, the problem is defined as *the minimum makespan problem on uniform parallel machines* [15] which is NP-hard. It is not practical to design a task scheduling system that can achieve the optimal scheduling due to the high complexity of searching for the solution.

5.5 A Practical Polynomial Time Solution

In order to find a good task allocation within a reasonable polynomial time, we relax the conditions of Theorem 5.1 and propose a heuristically solution for the task allocation problem in AMC, as shown in Fig. 5.5.

While Fig. 5.5 shows the way of allocating tasks in AMC from high level, Table 5.2 shows how the m independent tasks are divided into k groups that are allocated to k c-groups in polynomial time. In this solution, the m independent tasks are sorted in descending order of their execution time on the fastest core (any core in G_1). If the fastest core needs longer time to execute a task a than another task b, the workload of a is heavier than the workload of b. Based on the sorted tasks, we choose p_1, ..., p_{k-1} to divide the m tasks into k groups that are allocated to the k c-groups (i.e., G_1, ..., G_k) according to Algorithm 6.

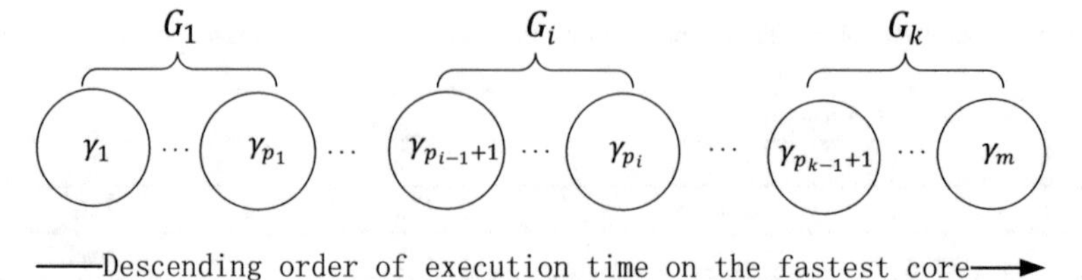

Fig. 5.5 Allocate m tasks with different workloads to k c-groups

Table 5.2 Allocate m tasks with different workloads to k c-groups

Tasks \ C-groups	G_1	G_2	...	G_i	...	G_k
γ_1	t_{11}	t_{12}	...	t_{1i}	...	t_{1k}
...	...	...	...	...	...	...
γ_{p_1}	$t_{(p_1)1}$	$t_{(p_1)2}$	...	$t_{(p_1)i}$	...	$t_{(p_1)k}$
γ_{p_1+1}	$t_{(p_1+1)1}$	$t_{(p_1+1)2}$	...	$t_{(p_1+1)i}$	...	$t_{(p_1+1)k}$
...	...	...	...	...	...	...
γ_{p_2}	$t_{(p_2)1}$	$t_{(p_2)2}$	...	$t_{(p_2)i}$	...	$t_{(p_2)k}$
...	...	...	...	...	...	...
$\gamma_{p_{i-1}+1}$	$t_{(p_{i-1}+1)1}$	$t_{(p_{i-1}+1)2}$	...	$t_{(p_{i-1}+1)i}$	...	$t_{(p_{i-1}+1)k}$
...	...	...	...	...	...	...
γ_{p_i}	$t_{(p_i)1}$	$t_{(p_i)2}$	...	$t_{(p_i)i}$	...	$t_{(p_i)k}$
...	...	...	...	...	...	...
$\gamma_{p_{k-1}+1}$	$t_{(p_{k-1}+1)1}$	$t_{(p_{k-1}+1)2}$	...	$t_{(p_{k-1}+1)i}$	...	$t_{(p_{k-1}+1)k}$
...	...	...	...	...	...	...
γ_m	t_{m1}	t_{m2}	...	t_{mi}	...	t_{mk}

We assume there are enough tasks to be allocated to the c-groups and each c-group will be allocated at least 1 task (i.e., $p_i < p_k$ if $i < k$). Observed from Table 5.2, once $p_1, ..., p_{k-1}$ are determined, the m tasks are divided into k groups. Because the values of $p_1, ..., p_{k-1}$ could be 1, 2, ..., m-1, the number of possible choices of dividing the m tasks equals to the number of possible choices of selecting $k-1$ numbers from 1, 2, ..., m-1. The above problem is a combination problem and the overall number of choices is C_{m-1}^{k-1}. Therefore, in Algorithm 8, we compare the estimated makespan of the tasks for each of all the C_{m-1}^{k-1} combinations of $p_1, ..., p_{k-1}$ and choose the combination of $p_1, ..., p_{k-1}$ that result in the minimum makespan.

C_{m-1}^{k-1} could be very large if both m and k are large. In the worst case, Algorithm 6 is of an exponential time complexity. Fortunately, AMC architectures only have two types of cores in most cases (i.e., $k = 2$) [11, 22, 25]. If $k = 2$, $C_{m-1}^{k-1} = C_{m-1}^{1} = m-1$, which increases with the number of tasks linearly. In addition, by grouping tasks into task classes, AATS can further greatly reduce m and thus can further significantly reduce C_{m-1}^{k-1} (the details will be explained in Sect. 5.7). Because both m and k are small in AATS, the overhead of Algorithm 8 is negligible for real world applications and real AMC architectures.

Algorithm 8 Static near-optimal polynomial time task allocation.

Require: A set of tasks $\{\gamma_1, ..., \gamma_m\}$;
Require: The ETT of tasks on k c-groups: t[m][k]
Require: The numbers of cores in c-groups $G_1, ..., G_k$: $N_1, ..., N_k$
Ensure: p[k-1]: $\{p_1, ..., p_{k-1}\}$

1: AllocateTask()
2: {
3: int p[k-1], q[k-1]; //q[k-1] stores the to-be-evaluated combination of $p_1, ..., p_{k-1}$
4: int i=0, min_span = MAXMUM_INT ;
5: while (Not all the settings are evaluated)
6: {
7: Get a new setting from C_{m-1}^{k-1} possible settings and update q[k-1] ;
8: if (any of $\{ \frac{\sum_{n=0}^{q[0]} t[n][0]}{N_1}, ..., \frac{\sum_{n=q[k-2]+1}^{m-1} t[n][k-1]}{N_k} \}$ > min_span)
9: continue ;
10: else
11: min_span = max $\{ \frac{\sum_{n=0}^{q[0]} t[n][0]}{N_1}, ..., \frac{\sum_{n=q[k-2]+1}^{m-1} t[n][k-1]}{N_k} \}$;
12: Copy q[k-1] to p[k-1] ;
13: }
14: return p[k-1] ;
15: }

In the above heuristically near-optimal solution, we assume the execution time table of the program has been constructed and all the items in the table are known. However, in real parallel applications, this assumption is not valid because these information is not known until they complete. How to apply the above theoretical solution to parallel programming environments is a challenging issue.

5.6 Design of Asymmetric-Aware Task Scheduling

In order to find a practical way to apply the above polynomial time solution in real-system scenario, in this section, we present an *Asymmetric-Aware Task Scheduler* (AATS) that leverages a static *history-based task allocation policy* and a dynamic *preference-based work-stealing policy*.

Figure 5.6 shows the processing flow of an application with AATS on an AMC architecture.

As shown in Fig. 5.6, using the history-based task allocation policy, AATS initially allocates tasks to the right c-groups statically. Basically we use history to predict the execution time of future tasks on cores in different c-groups. Tasks are classified into *task classes* according to their function names. Instead of allocating tasks directly, we allocate the task classes to different c-groups. For the same function f, we can collect the average execution time of the f-named tasks on cores in every c-group respectively in the history. Because the average execution time of each task class in every c-group is known from history, we can adopt Algorithm 8 to allocate the

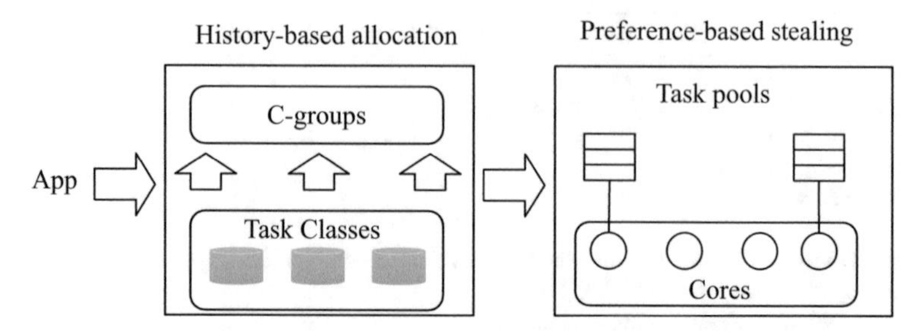

Fig. 5.6 The processing flow of an application with AATS on an AMC architecture

functions to different c-groups. Based on this allocation, tasks will be allocated to the c-group where its function name is allocated. If history workloads of tasks can reasonable reflect future workloads of tasks that execute the same function, this task allocation scheme will work well.

It is possible that in seldom cases where history cannot precisely predict the future, the allocation suggested by the above *history-based task allocation* is only an approximation of the optimal allocation. In order to further balance the workload, AATS adopts the *preference-based work-stealing policy* to further adjust the workloads dynamically among different c-groups.

The philosophy behind AATS is based on our theoretical analysis in Sect. 5.4: an optimal task allocation is more crucial to the makespan of parallel tasks than the rescuing policies like task snatching or stealing; and a workload-aware task snatching/stealing is better than random snatching/stealing. The history-based task allocation policy and the preference-based work-stealing policy are used to fulfill the philosophy.

5.6.1 Processing Flow of AATS

Without loss of generality, we assume the asymmetric cores in AMC can be divided into k c-groups G_1, ..., G_k, where G_i has N_i cores, and the cores in G_i are faster than the cores in G_j if $i < j$.

As presented before, instead of allocating the dynamically spawned tasks, AATS allocates the task classes to different c-groups. In order to support the strategy, AATS creates one task pool for each task class to store its tasks. When a task γ with a function name f is generated, its task pool is checked first. If the task pool for f-named tasks exists, γ is pushed into the correspondence task pool. If there is no task pool for f-named tasks, then a new task pool is created and γ is pushed into the new task pool.

Generally, we use a data structure $TC(f, ipc, n_1, ..., n_k, t_1, ..., t_k)$ to represent a task class, where f is the function name, *ipc* is the IPC (*Instruction-Per-Cycle*) of a task in the task class on a core in the fastest c-group G_1, n_i $(1 \leq i \leq k)$ is the number of tasks executed by cores in c-group G_i in history and t_i ($1 \le i \le k$) is the estimated execution time of a task in the task class on a core in c-group G_i. Note that, in any task class $TC(f, ipc, n_1, ..., n_k, t_1, ..., t_k)$, for any $1 \le i \le k$, n_i and t_i cannot be obtained directly because the tasks are spawned dynamically and their real execution time on different cores cannot be obtained until they are completed.

In order to allocate task classes to c-groups appropriately, the key issue is to obtain ipc, n_i ($1 \le i \le k$) and t_i ($1 \le i \le k$) of all the task classes precisely. Targeting this issue, AATS collects ipc, n_i ($1 \le i \le k$) and t_i ($1 \le i \le k$) of all the task classes based on historical statistics. Once the information of all task classes are determined, AATS uses history-based task allocation policy to allocate the task classes to different c-groups near-optimally using Algorithm 8.

After the task classes are allocated, AATS leverages the preference-based work-stealing policy to balance tasks among cores in the same c-group and among different c-groups dynamically. In the preference-based work-stealing policy, once cores in one c-group finish all the tasks allocated to the c-group, the cores help other c-groups to execute their tasks. To achieve this purpose, each core is given a *preference list* of task clusters (to be defined shortly). An idle core obtains a task according to the order of its preference list.

In the following two sections, we explain the details of the history-based task allocation policy and the preference-based work-stealing policy, respectively.

5.7 History-Based Task Allocation

During the execution of a parallel program, AATS assumes that tasks executing the same function in the current run have similar workloads. As for the assumption, although a function may show divergent behaviors depending on the inputs, the inputs of the tasks in the same task class in one run are often similar due to data parallelism. Empirically, in order to parallelize a serial program, the whole data set of the serial program is often been divided into many equal-size data blocks and each task will work on a single data block. Therefore, most well-designed data parallel programs obey this assumption. For example, in pipeline programs (e.g., Dedup and Ferret in Parsec [3]), tasks in different stages run in parallel. Tasks in the same stage execute the same function and they have similar workloads, but tasks in different stages execute different functions and they have different workloads. For programs that do not obey this assumption (such as divide-and-conquer programs), traditional work-stealing policy will be used to schedule the programs. Based on the above assumption, AATS uses the historical statistics collected during the execution of a program to estimate the workloads of future tasks in the same run. AATS collects the execution time and IPC of all the completed tasks using hardware performance counters.

Remember that AATS allocates task classes instead of tasks to the c-groups. Therefore, the first step in the history-based task allocation is building the task classes for the target parallel program. AATS builds task classes for non-batch programs and batch programs in different ways.

5.7.1 *Build Task Classes*

5.7.1.1 Non-batch Programs

In parallel programs whose tasks are not processed in batches, such as pipeline programs, the parallel tasks are generated dynamically at runtime. Because the tasks are spawned continually, it is not practical to group the un-processed tasks and determine the appropriate allocation directly. In this scenario, the task classes are created totally based on the collected historical information.

Therefore, for non-batch programs, AATS further assumes that the percentage of tasks executing the same function among all tasks is almost the same during the execution of a parallel application. As for this assumption, in many signal processing programs, different signals are input into the programs at a constant rate, where tasks processing different signals are created at a constant rate. As another example, most pipeline programs also obey this assumption. In pipeline programs, the data are divided into many data chunks and the data processing are divided into several stages. Because a task is launched for a data chunk at every stage and every data chunk needs go through all the stages, the assumption is approved.

Under this assumption, the near-optimal task allocation for the completed tasks are also near optimal for the future tasks. Therefore, in this case, the history-based core allocator searches the near-optimal task allocation for the completed tasks instead and then allocates the newly spawned tasks in the same allocation strategy. To find the appropriate allocation for the completed tasks, the tasks completed in history are also organized as task classes according to their function names.

We still use $TC(f, ipc, n_1, ..., n_k, t_1, ..., t_k)$, to represent a task class that is comprised of the completed tasks. In AATS, the task classes of completed tasks are updated in a timely manner. Once a task γ is completed, AATS collects its execution time and its IPC. Suppose the execution time and IPC of γ are t_γ and ipc_γ respectively. If task γ is executed by a core in the c-group G_1, then its task class $TC(f, ipc, n_1, ..., n_k, t_1, ..., t_k)$ is updated to Eq. 5.3.

$$TC(f, \frac{ipc \times n_1 + ipc_\gamma}{n_1 + 1}, n_1 + 1, ..., n_k, \frac{t_1 \times n_1 + t_\gamma}{n_1 + 1}, ..., t_k) \tag{5.3}$$

Otherwise, if γ is executed by a core in G_i ($i > 1$), n_i in its task class is updated to $n_i + 1$, and t_i in its task class is updated to $\frac{t_i \times n_i + t_1 \times \frac{ipc}{ipc_\gamma}}{n_i + 1}$ as described in Eq. 5.4. If there is no such a class, a new task class $TC(f, ipc, n_1, ..., n_k, t_1, ..., t_k)$ is created for f. In the newly created task class, $n_i = 1$, $t_i = t_\gamma$ and all the other items are 0.

$$TC(f, \frac{ipc \times n_1 + ipc_\gamma}{n_1 + 1}, n_1, ..., n_i + 1, ..., n_k, t_1, ..., \frac{t_i \times n_i + t_1 \times \frac{ipc}{ipc_\gamma}}{n_i + 1}, ..., t_k) \tag{5.4}$$

Note that, for task γ that is executed by a core in c-group G_i, AATS does not use its real execution time t_γ to update t_i in its task class $TC(f, ipc, n, t_1, ..., t_i, ..., t_k)$ but using its IPC ipc_γ to update t_i. This is because the execution time of a task on a core may increase due to events that the runtime system cannot control. This could be the case of the execution of an interrupt handler on the core where the task was scheduled, page-fault processing, or any bottom-half processing the OS does (load balancing or I/O related operations). Fortunately, we can use $t_1 \times \frac{ipc}{ipc_\gamma}$ to calculate the execution time of γ when it is not interrupted by any other events. Using the expected execution time of γ, the information in the task classes is accurate enough for construct DT of the correspondence program.

5.7.1.2 Batch Programs

Different from the non-batch programs, in batch programs, the parallel tasks are launched and processed in batches. Only when all the tasks in a batch are completed, the program launches another batch of tasks. For batch programs, the additional assumption for non-batch programs can be removed because AATS can obtain the real numbers of tasks in every task class.

If a batch program enters a new batch, in AATS, the workers do not execute the tasks immediately but letting the program generate all the tasks in the batch first. Because the tasks are divided into task classes and are distributed to different task pools when they are generated, once all the tasks in the batch are spawned, AATS gets many task classes of un-executed tasks and the correspondence task pools.

AATS uses a simplified data structure $TC(f, ipc, n, t_1, ..., t_k)$ to represent a task class in a batch program, in which n is the number of f-named tasks in the current batch and t_i $(1 \leq i \leq k)$ is the estimated execution time of the tasks in the task class on a core in c-group G_i. In $TC(f, ipc, n, t_1, ..., t_k)$, n can be collected by counting the number of tasks in the correspondence task pool. Based on the historical statistics, we calculate t_j $(1 \leq j \leq k)$ in $TC(f, ipc, n, t_1, ..., t_k)$. Let r represent the number of f-named tasks completed by cores in c-group G_i in history and let $ipc_{j1}, ..., ipc_{jr}$ represent their IPCs. We can calculate t_j in $TC(f, ipc, n, t_1, ..., t_j, ..., t_k)$ in Eq. 5.5.

$$t_j = t_1 \times \frac{ipc}{\sum_{m=1}^{r} ipc_{jm}/r} \tag{5.5}$$

Based on the information about the task classes, the next step is to allocate the task classes of the completed tasks to the k c-groups as described in Sect. 5.7.2.

5.7.2 Allocate Task Classes to C-Groups

Without loss of generality, we suppose there are overall m task classes in a parallel program. If the parallel program is not a batch program, the m task classes are denoted by $TC_1(f_1, ipc_1, n_{11}..., n_{1k}, t_{11}, ..., t_{1k})$, ..., $TC_m(f_m, ipc_m, n_{m1}..., n_{mk}, t_{m1}, ..., t_{mk})$. Otherwise, if the parallel program is a batch program, the m task classes are denoted by $TC_1(f_1, ipc_1, n_1, t_{11}, ..., t_{1k})$, ..., $TC_m(f_m, ipc_m, n_m, t_{m1}, ..., t_{mk})$. Note that, the m task classes TC_1, ..., TC_m are sorted in the descending order of t_{i1} $(1 \le i \le m)$.

If the task classes are ready, we apply Algorithm 8 to divide the task classes into k groups and allocate them to the k c-groups accordingly. In order to apply Algorithm 8, we need to build the duration table for the m task classes first. Tables 5.3 and 5.4 give the duration table for a non-batch program and a batch program respectively.

Similar to Table 5.1, in Tables 5.3 and 5.4, the very item at row TC_i and column G_a represent the time needed by a core in c-group G_a to execute all the tasks in task class TC_i. For non-batch program, because there are $\sum_{j=1}^{k} n_{ij}$ tasks in task class TC_i are completed in history and the expected execution time of a task in TC_i on a core in c-group G_a is t_{ia}, the very item at row TC_i and column G_a of Table 5.3 is $\sum_{j=1}^{k} n_{ij} \cdot t_{ia}$. For batch program, because there are overall n_i tasks in task class TC_i are generated in the current batch and the expected execution time of a task in TC_i on a core in c-group G_a is t_{ia}, the very item in row TC_i and column G_a of Table 5.4 is $n_i \cdot t_{ia}$.

Based on Tables 5.3 and 5.4, we apply Algorithm 8 to divide the task classes into k groups and allocate them to the k c-groups accordingly. We call the k groups of task classes *task clusters*. Since task clusters and c-groups are a one-to-one mapping, for the sake of convenience, we use G_i to represent both a task cluster and a c-group in the following discussion. Figure 5.7 illustrates how the history-based task allocation policy works. In the figure, task pool P_j $(1 \le j \le m)$ stores tasks in task class TC_j.

As mentioned before, in both Tables 5.3 and 5.4, the m task classes are sorted in the descending order of t_{i1} $(1 \le i \le m)$ instead of $\sum_{j=1}^{k} n_{ij} \cdot t_{i1}$ $(1 \le i \le m)$ or $n_i \cdot t_{i1}$ $(1 \le i \le m)$. In this way, if we unfold the task classes into tasks, the tasks in the new duration table are sorted in the descending order of their execution time on

Table 5.3 Duration table of a non-batch program with m task classes on an AMC with k c-groups

Task classes	C-groups				
	G_1	...	G_a	...	G_k
TC_1	$\sum_{j=1}^{k} n_{1j} \cdot t_{11}$	...	$\sum_{j=1}^{k} n_{1j} \cdot t_{1a}$	...	$\sum_{j=1}^{k} n_{1j} \cdot t_{1k}$
TC_2	$\sum_{j=1}^{k} n_{2j} \cdot t_{21}$	...	$\sum_{j=1}^{k} n_{2j} \cdot t_{2a}$	...	$\sum_{j=1}^{k} n_{2j} \cdot t_{2k}$
...	...	...	...	...	...
TC_i	$\sum_{j=1}^{k} n_{ij} \cdot t_{i1}$	...	$\sum_{j=1}^{k} n_{ij} \cdot t_{ia}$	...	$\sum_{j=1}^{k} n_{ij} \cdot t_{ik}$
...	...	...	...	...	...
TC_m	$\sum_{j=1}^{k} n_{mj} \cdot t_{m1}$	...	$\sum_{j=1}^{k} n_{mj} \cdot t_{ma}$	...	$\sum_{j=1}^{k} n_{mj} \cdot t_{mk}$

Table 5.4 Duration table of a batch program with m task classes on an AMC with k c-groups

Task classes	C-groups				
	G_1	...	G_a	...	G_k
TC_1	$n_1 \cdot t_{11}$	...	$n_1 \cdot t_{1a}$	...	$n_1 \cdot t_{1k}$
TC_2	$n_2 \cdot t_{21}$	...	$n_2 \cdot t_{2a}$	...	$n_2 \cdot t_{2k}$
...	...	...	...	...	...
TC_i	$n_i \cdot t_{i1}$	...	$n_i \cdot t_{ia}$	...	$n_i \cdot t_{ik}$
...	...	...	...	...	...
TC_m	$n_m \cdot t_{m1}$	...	$n_m \cdot t_{ma}$	...	$n_m \cdot t_{mk}$

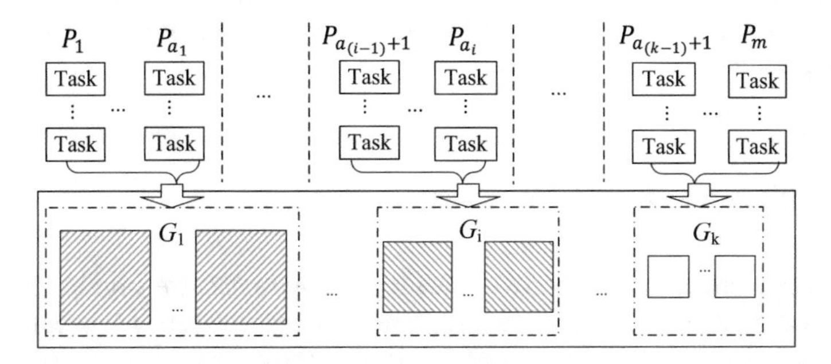

Fig. 5.7 Allocating m task classes to k c-groups in the history-based task allocation

the fastest core as in Table 5.1, which is the basis of Algorithm 8. However, if we unfold the task classes in Tables 5.3 and 5.4 into tasks, the new duration tables will have $M = \sum_{a=1}^{m} \sum_{b=1}^{k} n_{ab}$ rows and $M = \sum_{a=1}^{m} n_a$ rows respectively. In this case, Algorithm 8 has to check C_{M-1}^{k-1} possible combinations of p_1, ..., p_{k-1} to search for the optimal allocation of tasks to c-groups.

On the other hand, if the tasks are grouped into task classes as in Tables 5.3 and 5.4, Algorithm 8 only needs to check C_{m-1}^{k-1} possible combinations. Because a parallel program often has a great amount of tasks but only have a small number of task classes, $m \ll M$ and $C_{m-1}^{k-1} \ll C_{M-1}^{k-1}$. Essentially, by grouping tasks into task classes, we make sure that the tasks executing the same function are allocated to the same c-group. In this way, we can greatly reduce the tries needed to search for the appropriate allocation of tasks to c-groups. For instance, suppose 200 tasks that can be classified into 10 task classes are completed. If the program runs on an AMC with four types of cores (normally, there are only two types of cores in an AMC), the number of combinations of p_1, ..., p_{k-1} is reduced from $C_{199}^{3} = 1,293,699$ to $C_9^3 = 84$ by grouping tasks into task classes. Therefore, the overhead of Algorithm 8 in AATS is small.

It is worth noting that all the information required by thehistory-based task allocation is collected automatically. The number of cores in every c-group can be acquired from the operating system. The execution time and IPC of a task are acquired at runtime. Once a task is completed, the information about its task class is updated as presented in Sect. 5.7.1.

In addition, the execution time of a memory-intensive task can vary from run to run due to the contention on shared resources. The contention may result in slightly unbalanced workload in AATS. For a newly-generated task, since we use the IPCs of all the tasks completed in history in its task class to estimate its execution time, the calculated execution time is close to its real execution time. Even the contention on shared resources incurs slightly load unbalancing, AATS can use the preference-based work-stealing policy to further balance the workload dynamically by scheduling the tasks at runtime.

5.8 Preference-Based Work-Stealing

AATS uses a preference-based work-stealing policy to balance workloads dynamically. In our situation, task scheduling is complex since there are overall m task pools, labeled as P_1, ..., P_m, corresponding to the m task classes TC_1, ..., TC_m. AATS first tries to balance the workloads within the same c-group before scheduling tasks across c-groups. We use a core c from the c-group G_i in Fig. 5.7 as an example to explain the details of the preference-based work-stealing policy.

5.8.1 *Scheduling Within a C-Group*

If core c from c-group G_i is free, it first tries to obtain tasks from the task pools that store tasks allocated to its c-group G_i (i.e., $P_{a_{(i-1)}+1}$, ..., P_{a_i} as shown in Fig. 5.7). Since multiple task pools are allocated to G_i, c needs to decide to obtain a task from which task pool first.

A basic strategy for choosing victim task pool to obtain task is choosing task pools in the order of $P_{a_{(i-1)}+1}$, ..., P_{a_i} that is the order of task classes in Fig. 5.7. Only when the task pool P_j ($a_{(i-1)} + 1 \leq j < a_i$) is empty, c tries to obtain a task from the next task pool P_{j+1} until it gets a task.

This basic strategy is similar to work-sharing strategy in which all the cores share a single task pool. In the basic strategy, cores in G_i try to execute all the tasks in one task pool before moving to the next task pool allocated to G_i. Therefore, the basic strategy often causes serious lock contention on the task pools similar to work-sharing since many cores in G_i may try to lock the same task pool for obtaining new tasks. The serious lock contention may degrade the performance of AATS.

To relieve the lock contention, we decide to use a strategy borrowed from random work-stealing that has been proved to be effective. In traditional random work-stealing, each core has a task pool. When a core is free, a core randomly chooses a victim core and tries to steal a task from the victim core's task pool. In our scenario, there are multiple task pools as well but they are associated with task classes not cores. In AATS, if c is free, it randomly chooses a task pool P_j from $P_{a_{(i-1)}+1}, ..., P_{a_i}$ and tries to obtain a task from P_j. If P_j is empty, it randomly chooses another task pool and tries to obtain a task from the new chosen task pool until c gets a task. In this case, since there are multiple task pools for obtaining tasks, the lock contention is much lower and the performance would be better.

5.8.2 *Scheduling Among C-Groups*

If all the task pools allocated to G_i are empty, which means all the tasks allocated to G_i complete, AATS allows c to execute tasks allocated to other c-groups in order to balance the workloads among different c-groups dynamically. The complexity arises when deciding which c-group to choose in this situation. The following preference-based work-stealing strategy gives our solution. In the preference-based work-stealing policy, each core is given a *preference list* of task clusters. Each task cluster contains multiple task pools. The preference list of a core contains all the k task clusters that are ordered as detailed below.

Algorithm 9 Preference-based task scheduling

Require: A core c from the c-group G_i
Require: c's preference list $\{G_i, ..., G_k, G_{i-1} ..., G_1\}$

```
1: ObtainNewTask()
2: {
3:   while (c has not obtained a task)
4:   {
5:     for (each G_j ∈ {G_i, ..., G_k, G_{i-1}..., G_1})
6:     {
7:       while (not all the task pools allocated to G_j are empty)
8:       {
9:         c randomly chooses a task pool P_a allocated to G_j ;
10:         c tries to obtain a task t from P_a ;
11:         if (succeed) {
12:           return t ;
13:         }
14:       }
15:     }
16:   }
17: }
```

For core c in the c-group G_i, its preference list is created as $\{G_i, G_{i+1}, ..., G_k, G_{i-1}, G_{i-2}, ..., G_1\}$ as shown in Fig. 5.8. We design a *help-the-weaker-first* principle to guide the generation of the preference list in the figure. This principle can help reduce the makespan. For example, if a core obtains a task that is allocated to faster cores, it needs a long time to execute the obtained task, which may prolong the makespan. On the contrary, if a core obtains a task that is allocated to slower cores, it can execute the obtained task in a shorter time and relieve the pressure on slow cores. However, this preference list does not prevent slow cores to obtain tasks from fast cores. When the slow cores have no tasks to execute, they can obtain tasks from the busy fast cores.

Once c decides to help which c-group G_j ($1 \leq j \leq k$), it randomly selects the victim task pool from the task pools allocated to G_j following the same strategy described in Sect. 5.8.1. Algorithm 9 shows the preference-based task scheduling algorithm adopted by each core for obtaining a new task.

Figure 5.9 shows an example runtime structure of AATS on an asymmetric quad-core architecture with three different types of cores. That is, there are three c-groups G_1 (with core c_0), G_2 (with c_1 and c_2) and G_3 (with c_3).

Therefore, task classes are classified into 3 task clusters (G_1, G_2 and G_3) accordingly. Table 5.5 shows the corresponding preference lists of G_1, G_2 and G_3. The preference lists of the four cores are generated based on the *help-the-weaker-first* principle in Fig. 5.8. For example, c_3 will always look for tasks from the G_3 pools first, which have the tasks that are allocated to c_3's c-group using the history-based task allocator. Then it will search the G_2 pools, and finally the G_1 pools.

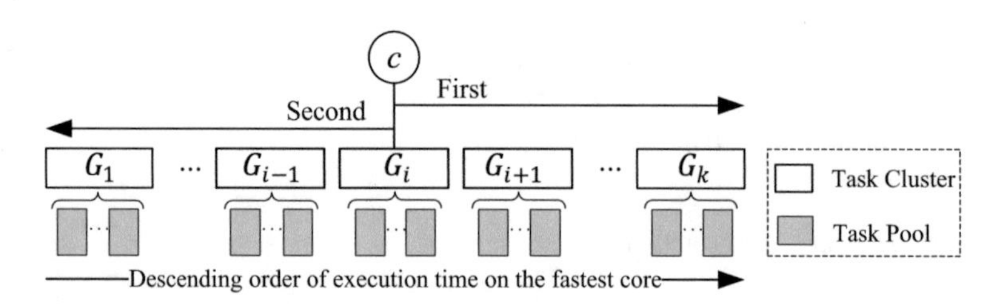

Fig. 5.8 Preference list of the core c in the c-group G_i

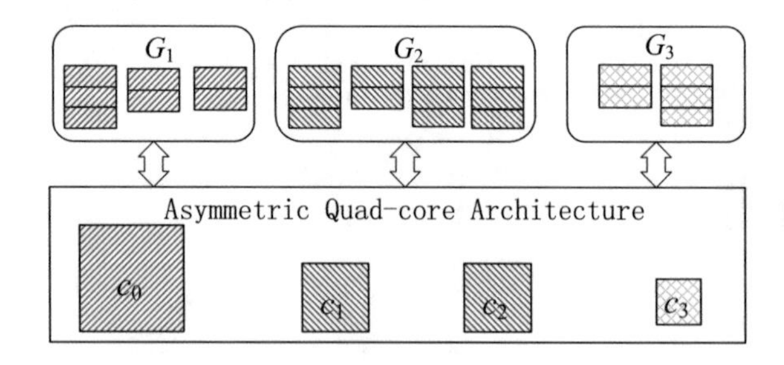

Fig. 5.9 An example runtime structure of AATS

Table 5.5 The corresponding preference lists of the c-groups in Fig. 5.9

C-group	Core	Preference list
G_1	c_0	$\{G_1, G_2, G_3\}$
G_2	c_1 and c_2	$\{G_2, G_3, G_1\}$
G_3	c_3	$\{G_3, G_2, G_1\}$

5.9 Implementation Methodology of AATS

AATS has been implemented by modifying MIT Cilk. MIT Cilk is one of the earliest open source parallel programming environments that implement work-stealing [8].

We have ported MIT Cilk to support the preference-based work-stealing policy. To help task classification, we have modified *cilk2c* to record a task's function name in the task frame. When a new task is spawned, it is subsumed into its task class and pushed into the corresponding task pool according to its function name stored in the task frame. In AATS, each worker tracks the execution time of the tasks executed by it. Once a task is completed, the worker updates the information of the correspondence task class.

AATS launches a helper thread that implements the history-based task allocation policy at runtime. The helper thread periodically (e.g., every 1ms) checks every core to find out if it has completed some tasks. Once there is a completed task, the helper thread updates the information of the correspondence task class. The helper thread is scheduled by the OS to any free core at runtime. Our experimental results show that the extra overhead incurred by the helper thread is very small.

We have ported Cilk to spawn tasks adopting the parent-first policy since AATS tends to generate all the tasks as soon as possible so that the history-based task allocator can allocate them to different c-groups in a short time. In addition, it is difficult to collect the workload information of tasks with the child-first policy. If a core is executing a task γ, with the child-first policy, it is very likely the core will also execute γ's child tasks before γ is completed. Therefore, γ's workload information may not be collected correctly as it could include the workloads of γ's child tasks. As a result, we have modified *cilk2c* to spawn tasks with the parent-first policy.

In order to construct the duration table of a program, for each task class, we need to collect the IPCs of its tasks on cores of all the c-groups using hardware performance counters. Based on the duration table, we can apply Algorithm 8 to allocate task classes to c-groups. If not all the items in the duration table are determined, the task classes cannot be allocated to different c-groups appropriately. To collect the information as soon as possible, motivated by random work-stealing strategy [8], any core c will grab a task from a random task pool when c is free. After all the task classes are built, the history-based task allocator can allocate task classes to c-groups and then AATS adopts preference-based task scheduler to balance the workloads dynamically.

An interesting detail of the AATS implementation is that AATS schedules the main task of a parallel program on the fastest core as in [21]. This is because the main task often has time-consuming serial initialization code before spawning tasks. If the main task is executed by a slow core, it will increase the makespan of the program. To exclude the impact of this optimization, we make all the other schedulers in the experiment section launch the main task on the fastest core, though those schedulers may launch the main task on a randomly chosen core. If the chosen core is slow, which is very likely, their performance will be even worse.

Not surprisingly, AATS has a limitation. If most tasks in an application execute the same function, the history-based task allocation algorithm can only identify a few task classes that cannot be evenly allocated to the c-groups. For example, recursive divide-and-conquer programs such as *nqueens* and *fib* are not suitable for AATS. To cope with this problem, we have modified the compiler *cilk2c* to check for the divide-and-conquer programs at compile time by analyzing the task generating pattern in the source code. If any function in the source code generates new tasks that run the same function as itself, the program is assumed to be a divide-and-conquer program. For divide-and-conquer programs, the task scheduling policies introduced in previous chapters can be used instead to schedule the program. Therefore, the above limitation will not affect the applicability of AATS since the compiler can identify the class of programs that are suitable for AATS.

5.10 Performance of AATS

In this section, we evaluate AATS, including its performance over the current task schedulers, the effectiveness of the preference-based work-stealing policy in AATS, and the scalability of AATS. After that, we discuss that whether we should integrate task snatching into AATS or not. Lastly, we discuss other issues related to AATS.

5.10.1 Experimental Configurations

We use a 16-core server that has four AMD Quad-core Opteron 8380 processors (codenamed "Shanghai") and a 32-core server that has four Intel Octal-core Xeon(R) X7560 processors to evaluate the performance of AATS. In the AMD Opteron 8380 processor, each core can run at 2.5, 1.8, 1.3 and 0.8GHz. In the Intel Xeon X7560 processor, each core can run at 11 different frequencies. We adjust the frequency of each core to emulate different single-ISA AMC architectures in the experiment. To emulate AMC architectures, we use all the four possible frequencies in the server built with AMD processors. We use 2.262, 1.862, 1.463 and 1.064 GHz in the server built with Intel processors.

Figure 5.10 provides the topology of the emulated AMC architectures. We use A-*a*-*b*-*c*-*d* to represent the emulated AMC architecture on AMD server that has *a* cores

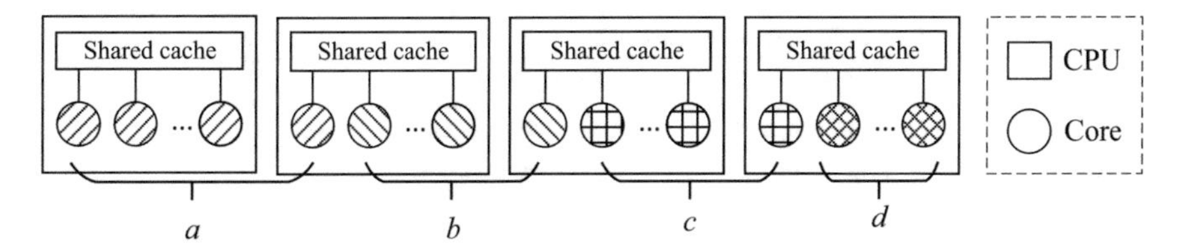

Fig. 5.10 Topology of the emulated AMC architectures. The numbers of cores running at four different frequencies are *a*, *b*, *c* and *d* respectively

Table 5.6 The emulated AMC architectures in the experiment

AMD-based server	2.5 GHz	1.8 GHz	1.3 GHz	0.8 GHz
A-2-2-2-10	2	2	2	10
A-4-4-4-4	4	4	4	4
A-2-0-0-14	2	0	0	14
A-4-0-0-12	4	0	0	12
A-8-0-0-8	8	0	0	8
A-12-0-0-4	12	0	0	4
A-16-0-0-0	16	0	0	0
Intel-based server	2.262 GHz	1.862 GHz	1.463 GHz	1.064 GHz
I-8-0-0-24	8	0	0	24
I-8-8-8-8	8	8	8	8
I-16-0-0-16	16	0	0	16
I-24-0-0-8	24	0	0	8
I-32-0-0-0	32	0	0	0

running at 2.5GHz, *b* cores running at 1.8GHz, *c* cores running at 1.3 GHz and *d* cores running at 0.8GHz. Similarly, we use I-*a*-*b*-*c*-*d* to represent the emulated AMC architecture on Intel server that has *a* cores running at 2.262GHz, *b* cores running at 1.862GHz, *c* cores running at 1.463 GHz and *d* cores running at 1.064GHz.

Note that, Intel processors do not support per-core DVFS but only support per-CPU DVFS. Table 5.6 lists the emulated AMC architectures.

Because AATS is proposed to improve the performance of both batch and non-batch applications with tasks that have different workloads, we use benchmarks listed in Table 5.7, to evaluate the performance of AATS.

The source code of benchmarks are from their official websites [16] but adapted to run on MIT Cilk. In the batch-based benchmarks, the program launches different numbers of independent tasks (more than 128 tasks on AMD-based server and 256 tasks on Intel-based server) in different batches. In these benchmarks, tasks work on independent data sets of different sizes in parallel. For instance, in our configuration, *LZW* compresses more than 128 files of different sizes in parallel. Because the size of data set for different tasks are different, the workloads of tasks are different as well. In the pipeline-based benchmarks, the execution of a program has several parallel

Table 5.7 Benchmarks used in the experiments

Name	Type	Description
BWT	Batch-based	Burrows wheeler transform
DMC	Batch-based	Dynamic markov coding
GA	Batch-based	Island model of genetic algorithm [26]
LZW	Batch-based	Lempel-Ziv-Welch data compression
MD5	Batch-based	Message digest algorithm
SHA-1	Batch-based	SHA-1 cryptographic hash function
Dedup	Pipeline-based	Dedup from PARSEC benchmark suite [3]
Ferret	Pipeline-based	Ferret from PARSEC benchmark suite [3]

stages. Tasks in different stages run in parallel but communicate with each other via pipelines. For each test, every benchmark is run ten times. Because the execution time is quite stable, the average execution time is used as the result.

For pipeline-based benchmarks, AATS allocates tasks to c-groups adopting the method in Sect. 5.7.1.1. For batch-based benchmarks, AATS allocates tasks to c-groups adopting the method in Sect. 5.7.1.2.

We compare the performance of AATS with the performance of three other task schedulers: MIT Cilk, PFWS and RTS in AMC architectures. Although MIT Cilk is originally proposed to balance fine-grained tasks [4], the internal work-stealing strategy is still one of the most efficient dynamic load-balancing strategies to balance coarse-grained tasks, such as tasks in batch-based programs and pipeline-based programs [17–19].

In MIT Cilk (denoted as Cilk for short) [4], tasks are spawned with the child-first policy and scheduled with the traditional work-stealing policy. In PFWS (Parent-First Work-Stealing) [9], parallel tasks are spawned with the parent-first policy and scheduled with the traditional work-stealing policy. In RTS (Random Task-Snatching) [1], tasks are also spawned and scheduled as in Cilk, but a faster core snatches tasks from a randomly chosen slower core if the faster core cannot steal any task. The snatch operation is implemented by swapping the two threads on the faster core and the slower core.

To evaluate the effectiveness of the *help-the-weaker-first* policy,, we also compare the performance of AATS with AATS-NP, a scheduler that adopts the history-based task allocator but cores in one c-group are not allowed to obtain tasks that are allocated to other c-groups. In this way, AATS-NP is able to show only the performance of the history-based task allocator. To ensure fairness of comparison, AATS, PFWS, RTS and AATS-NP are implemented by modifying MIT Cilk.

5.10.2 Performance on Emulated Platform

We have tested the performance of the benchmarks in all the $6 + 4 = 10$ AMC architectures and the $1 + 1 = 2$ symmetric multi-core architectures. In the following experiments, the performance of a benchmark is normalized to its performance with the traditional random work-stealing scheduler, Cilk.

Figure 5.11 presents the performance of the batch-based benchmarks in all the emulated AMC architectures and Fig. 5.12 presents the performance of the pipeline-based benchmarks in all the emulated AMC architectures. In Figs. 5.11 and 5.12, the y axis on the left is the execution time of the benchmarks in AMC emulated on the

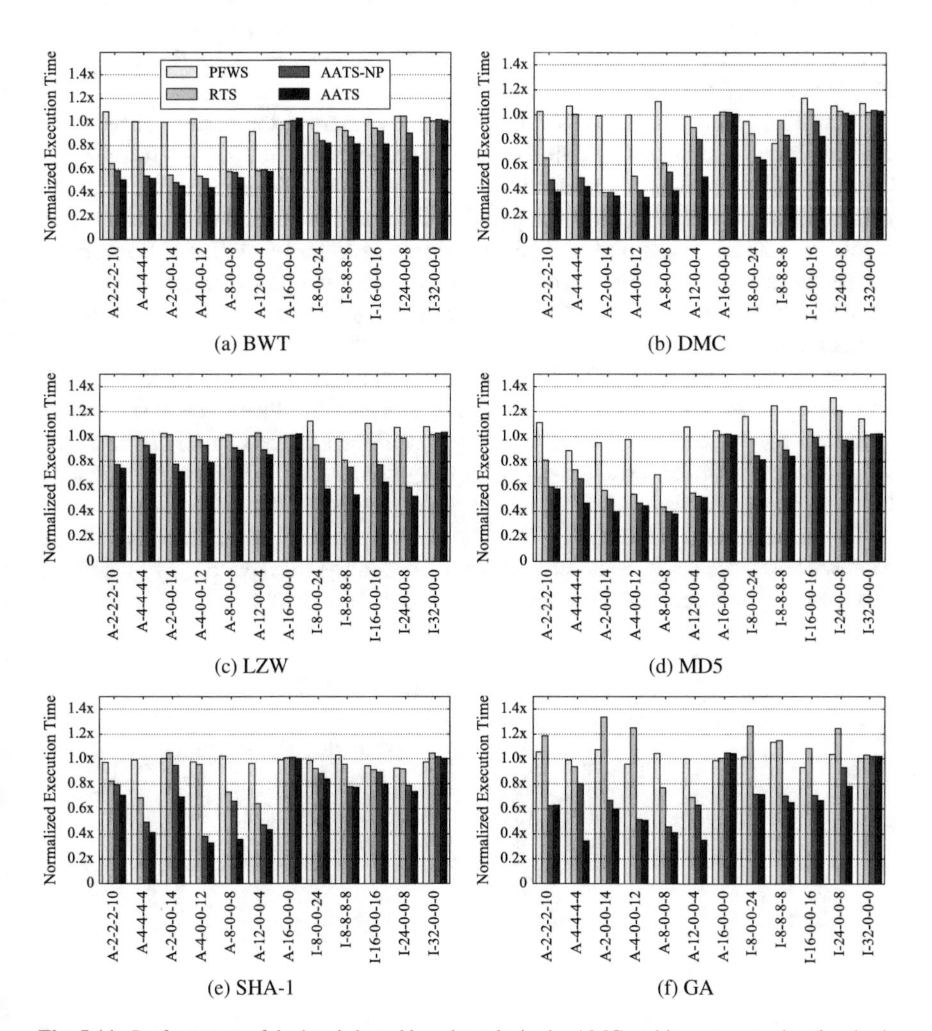

Fig. 5.11 Performance of the batch-based benchmarks in the AMC architectures emulated on both the AMD-based server and Intel-based server

AMD server and the y axis on the right is the execution time of the benchmarks in AMC emulated on the Intel server.

From Fig. 5.11 we can find that AATS can significantly improve the performance of the batch-based benchmarks in all the emulated AMC architectures. In A-2-2-2-10 to A-12-0-0-4 that are emulated on the AMD server, AATS improves the performance of batch-based benchmarks, with the performance gains ranging from 10.7 to 66.1% compared to Cilk and PFWS, and with performance gains ranging from 2.4 to 65.6% compared to RTS. In I-8-0-0-24 to I-24-0-0-8 that are emulated on the Intel server, AATS improves the performance of batch-based benchmarks, with the performance gains ranging from 3.3 to 38.1% compared to Cilk and PFWS, and with the performance gains ranging from 4.2 to 51.1% compared to RTS. For example, for SHA-1 in A-4-0-0-12 in Fig. 5.11e, AATS reduces the execution time up to 66.1% compared to Cilk. For LZW in I-8-8-8-8 in Fig. 5.11c, AATS reduces the execution time up to 38.1% compared to Cilk.

Figure 5.12 shows that AATS can significantly improve the performance of the pipeline-based benchmarks in all the AMC architectures. In A-2-2-2-10 to A-12-0-0-4, AATS improves the performance of pipeline-based benchmarks, with the performance gains up to 29.1% compared to Cilk and PFWS, and with performance gains up to 28.2% compared to RTS. In I-8-0-0-24 to I-24-0-0-8, AATS improves the performance of pipeline-based benchmarks, with the performance gains up to 44.9% compared to Cilk and PFWS, and with the performance gains up to 36.8% compared to RTS.

The good performance of AATS comes from its balanced workloads in the AMC architectures. With the history-based task allocator, AATS allocates tasks with heavy workload to fast cores and tasks with light workload to slow cores. Even if the workloads are not balanced as expected due to approximation, AATS can dynamically balance the workloads in AMC using the preference-based task scheduler.

On the contrary, in Cilk and PFWS, it is very likely that long tasks are scheduled to slow cores since tasks are stolen randomly. Scheduling a task with heavy workload to a slow core can seriously prolong the makespan of parallel tasks.

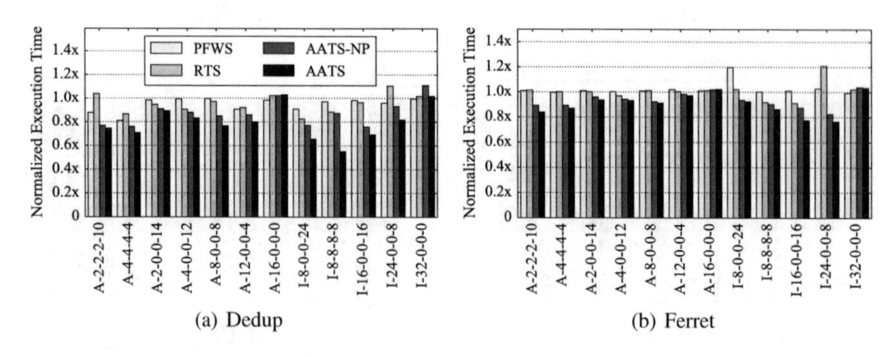

Fig. 5.12 Performance of the pipeline-based benchmarks in the AMC architectures emulated on both the AMD-based server and Intel-based server

From Fig. 5.11, we can find that AATS performs better for batch-based programs in the AMCs emulated on the AMD server than the AMCs emulated on the Intel server. The better performance on the AMD server comes from the large gap between the fast core speed and the slow core speed. As shown in Table 5.6, the speed of the slowest cores is only $\frac{0.8}{2.5} = 32\%$ of the speed of the fastest cores in A-2-2-2-10 to A-12-0-0-4 while the speed of the slowest cores is $\frac{1.064}{2.262} = 47\%$ of the speed of the fastest cores in I-8-0-0-24 to I-24-0-0-8. A task with heavy workload is slowed down by more times in A-2-2-2-10 to A-12-0-0-4 if the task is scheduled to a slow core. Therefore, the larger the gap between the fastest cores and the slowest cores in an AMC architecture is, the more AATS can improve the performance of applications that have tasks with different workloads.

Compared to Cilk and PFWS, RTS can also improve the performance of most benchmarks in AMC architectures. This is because in RTS faster cores can randomly snatch tasks from slower cores and the snatched tasks can be completed earlier, which can reduce the makespan of the parallel tasks. As a result, comparing to Cilk and PFWS, for most benchmarks, RTS improves the performance of the benchmarks up to 60.9% in A-2-2-2-10 to A-12-0-0-4 emulated on the AMD-based server and up to 35.6% in I-8-0-0-24 to I-24-0-0-8 emulated on the Intel-based server.

However, for many other benchmarks, such as GA in A-2-0-0-14 and MD5 in I-24-0-0-8, RTS even degrades the performance of the benchmarks due to the overheads that come from the frequent task snatching (or context switching). In addition, since RTS is not aware of the workloads of the tasks, it is possible for faster cores to snatch tasks with light workload, in which case the makespan cannot be reduced. Especially, in AMC emulated in Intel server, there are overall 32 cores. The large number of cores often lead to more task snatching operations in RTS. Therefore, RTS performs poor in I-8-0-0-24 to I-24-0-0-8 and it performs much worse than AATS.

For symmetric multi-core architectures, AATS schedules tasks in the similar way as PFWS. Therefore, as shown in Figs. 5.11 and 5.12, AATS performs the same as PFWS on A-16-0-0-0 and I-32-0-0-0. The overhead in AATS is negligible compared with traditional work-stealing in symmetric architecture.

Figures 5.11 and 5.12 also show that AATS can adapt to different AMC architectures and improve performance of benchmarks automatically. In addition, when an AMC architecture has a small number of fast cores (e.g., 8 fast cores in I-8-0-0-24, 2 fast cores in A-2-0-0-14), the frequent context switching on fast cores that comes from task snatching reduces the computing time of fast cores on tasks. In this case, RTS degrades the overall performance of some benchmarks (e.g., GA) compared with Cilk and PFWS.

5.10.3 Effectiveness of the Preference-Based Work-Stealing

Figures 5.11 and 5.12 also shows the performance of batch-based programs and the performance of pipeline-based programs with AATS-NP. As shown in the two figures, the performance of AATS is always better than the performance of AATS-NP. Especially, for Dedup in I-8-8-8-8, AATS-NP even prolongs the execution time of Dedup up to 32.3% compared to AATS. For GA in A-4-4-4-4, AATS-NP even prolongs the execution time of GA up to 45.8% compared to AATS. The preference-based work-stealing policy in AATS is very helpful when handling slightly unbalanced workloads. Since the history-based task allocation may mis-allocate the tasks to the wrong c-groups due to its static approximation of the workloads of dynamic tasks, the preference-based work-stealing can remedy this imprecision. To this end, we can conclude that the preference-based work-stealing works effectively in AMC architectures.

It is worth noting that the history-based task allocation has mostly done effective allocation of tasks according to Figs. 5.11 and 5.12. AATS-NP performs better than Cilk and PFWS, which means the history-based allocation algorithm is more effective than random task stealing in terms of load-balancing in AMC architecture.

In symmetric architectures (i.e., A-16-0-0-0 and I-32-0-0-0), AATS schedules tasks in a similar way to the pure parent-first work-stealing policy. From the figures, we can find that the benchmarks achieve similar performance in AATS, PFWS and Cilk in A-16-0-0-0 and I-32-0-0-0 that have symmetric cores respectively. Therefore, the overhead in AATS (mainly caused by the helper thread) is negligible compared with traditional work-stealing in symmetric architecture.

5.10.4 Scalability of AATS

Figure 5.13 evaluates the scalability of AATS. It gives the performance of GA under different distributions of workloads in A-8-0-0-8 and I-24-0-0-8, though other benchmarks show similar results in various AMC architectures. In the AMC emulated on AMD server that has 16 cores, GA launches 128 tasks with 4 different workloads in each batch. In the AMC emulated on Intel server that has 32 cores, GA launches 256 tasks with 4 different workloads in each batch. The number of tasks with each type of workload is adjusted to evaluate the scalability of AATS when the number of tasks with heavy workload increases. In Fig. 5.13a, the distribution of workloads from high to low follows the pattern α, α, α, $128 - 3\alpha$. In Fig. 5.13b, the distribution of workloads from high to low follows the pattern α, α, α, $256 - 3\alpha$. In both the two figures, α is adjusted as shown by the x-axis in the figures. In Fig. 5.13, the fastest core needs less than 200 μs to process the tasks with the lowest workload and needs less than 10 ms to process the tasks with the highest workload.

From the figure we can see that AATS works fine under different distributions of workloads. In A-8-0-0-8, when α is small and the workloads are mostly light, AATS reduces the GA execution time by 55.4% compared to Cilk. When α is large and the workloads are mostly heavy, AATS can still reduce the execution time by 17.2% compared to Cilk. In I-24-0-0-8, when α is small, AATS reduces the GA execution time by 22.8% compared to Cilk. When α is large and the workloads are

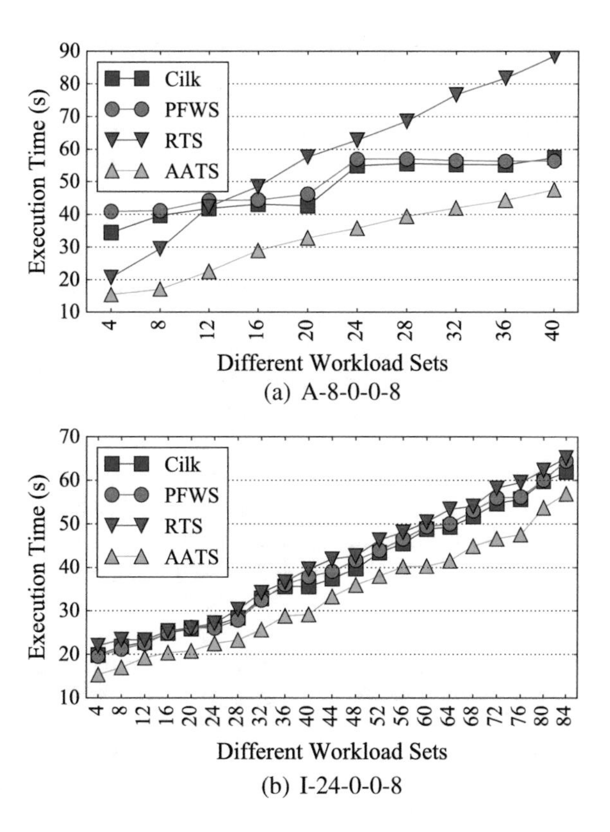

Fig. 5.13 Performance of GA with different workloads in A-8-0-0-8 and I-8-0-0-24 with Cilk, PFWS, RTS, and AATS

mostly heavy, AATS can still reduce the execution time by 8.1% compared to Cilk. Therefore, AATS is scalable with and can adapt to different workloads.

However, in A-8-0-0-8 as shown in Fig. 5.13a, RTS does not work well when the workloads are mostly heavy (e.g. α is 20), as it even degrades the performance of GA by 54.1% compared to Cilk and PFWS. This is because fast cores are not able to snatch all the heavy tasks that are allocated to the slow cores when there are too many heavy tasks. Moreover, the computing ability of fast cores is wasted at frequent context switching when the workloads are mostly heavy. This result again supports our philosophy of AATS that an optimal task allocation is more important than rescuing policies such as task snatching.

In I-24-0-0-8 as shown in Fig. 5.13b, RTS works even worse than Cilk and PFWS for all the workloads as it degrades the performance up to 10.7% compared to Cilk. In I-24-0-0-8, the reduced execution time of GA that origins from task snatching in RTS is small since the difference between the speed of fast cores and the speed of slow cores is small. It is quite possible that the reduced execution time is smaller than the increased execution time that origins from the context switching in RTS. Therefore, if the gap between the speed of fast cores and the speed of slow cores is small, the performance of RTS is poor.

5.10.5 Integrating Task-Snatching in AATS

It is also of interest to discover whether or not the task-snatching technique is also effective to AATS and thus should be integrated into AATS. To investigate this issue, we implemented a scheduler AATS-TS, where fast cores snatch tasks from slow cores when the fast cores cannot obtain any tasks using the preference-based task scheduling policy.

In AATS-TS, when a core intends to snatch a task, it selects a slower core with the largest task. In this way, large tasks that affect the makespan seriously can be snatched to fast cores and completed earlier. Therefore, our workload-aware snatching policy is better than the *random snatching* in RTS, as explained in Sect. 5.3.1. Moreover, workload-aware snatching causes fewer snatching operations than the random snatching, since randomly snatched small tasks take less time for the fast cores to complete, which causes the fast cores to snatch more often.

Figure 5.14 also shows the performance of the benchmarks in AATS and AATS-TS in A-4-0-0-12 and I-8-8-8-8. From the figure we see surprisingly that the performance of AATS-TS is slightly worse than AATS. Especially, for BWT and Ferret in A-4-0-

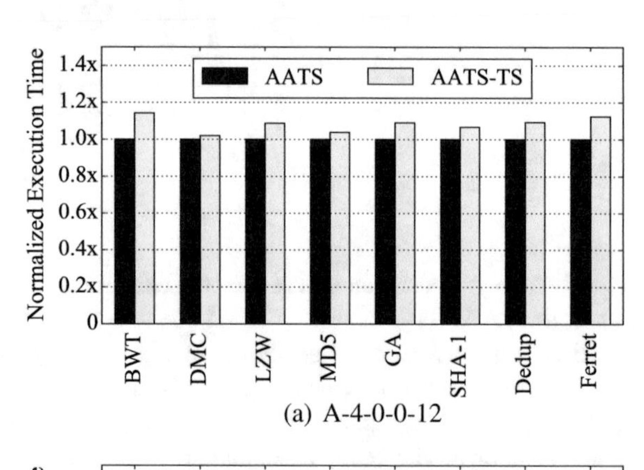

(a) A-4-0-0-12

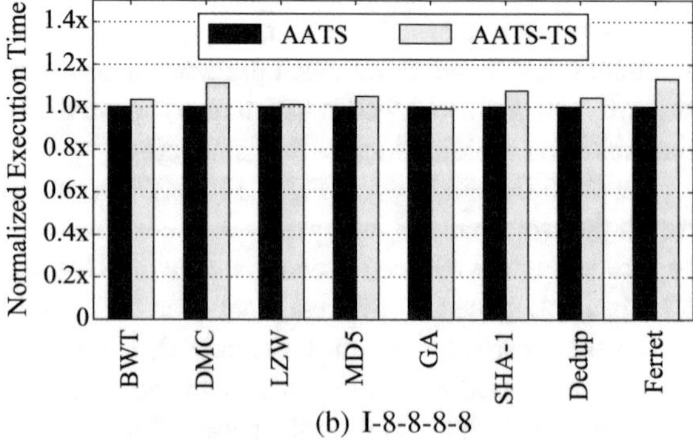

(b) I-8-8-8-8

Fig. 5.14 Performance of the benchmarks in AATS, AATS-OLD and AATS-TS

0-12 and DMC and Ferret in I-8-8-8-8, AATS-TS increases the execution time up to 14.2% compared to AATS.

Figure 5.14 tells us that AATS has satisfactorily balanced the workloads in AMC architectures. When the workloads are balanced among cores in AMC, it is not worthwhile to snatch tasks from slower cores since the slower cores are also close to completion. The extra overhead incurred by the snatching operations simply makes AATS-TS perform worse. Therefore, there is no need for AATS to adopt task-snatching.

5.11 Summary

Single-ISA AMCs are promising due to their high performance and power efficiency. It is essential for parallel applications to run on AMC architectures efficiently. Though task scheduling policies like work-stealing work efficiently for parallel applications in symmetric multi-core architectures, they cannot balance the workloads well in AMC since they have no knowledge of task workloads and schedule tasks randomly to the performance-asymmetric cores.

From our theoretical analysis, we know that the initial optimal task allocation is more crucial to the makespan than any rescuing means for a non-optimal allocation and that static task allocation can produce near-optimal allocation if the workloads of the tasks are known. Therefore, we propose history-based task allocation policy that takes advantage of the static allocation by using the historical statistics of the tasks to predict the execution time of future tasks on cores in different c-groups. From our experiments we showed that the history-based task allocator can produce appropriate allocation and its extra overhead is small.

For any occasional inaccurate or incorrect allocation of tasks, the preference-based work-stealing policy comes to play. It can remedy any slightly unbalanced allocation and effectively schedule tasks among c-groups. The experimental results show that AATS is effective and our approach to the scheduling problem in single-ISA AMC is valid.

In summary, AATS can significantly improve the performance of applications that consist of tasks with different workloads in all the emulated AMC architectures. Especially, AATS can adapt to various AMC architectures automatically and the performance of AATS is better when there are more fast cores in AMC. The experiment also justifies that both the history-based task allocation policy and the preference-based work-stealing policy in AATS are effective. In addition, although the task-snatching policy in RTS can improve the performance of applications independently, the integrating of task-snatching with AATS degrades the performance of AATS.

5.11.1 Chapter Highlights

The following highlights of this chapter could be of your interest:

- We have identified, defined, and formalized the problem of unbalanced workloads in AMC architectures and have given theoretical guidance to optimal task allocation in AMC architectures.
- We have proposed a history-based task allocation strategy that can allocate tasks in single-ISA AMC architectures near-optimally.
- We have proposed a novel preference-based work-stealing strategy that can effectively balance workloads among different groups of cores.
- Based on the above techniques, we have implemented AATS, which achieves a performance gain of up to 66.1% compared to the random work-stealing approach commonly employed.

References

1. M. Bender and M. Rabin, "Scheduling Cilk multithreaded parallel programs on processors of different speeds," in *Proceedings of the 12nd annual ACM Symposium on Parallel Algorithms and Architectures*. ACM, 2000, pp. 13–21.
2. M. Bhadauria and S. McKee, "An approach to resource-aware co-scheduling for cmps," in *Proceedings of the 24th ACM International Conference on Supercomputing*. ACM, 2010, pp. 189–199.
3. C. Bienia, S. Kumar, J. P. Singh, and K. Li. The parsec benchmark suite: characterization and architectural implications. In *Proceedings of the 17th international conference on Parallel architectures and compilation techniques*, pages 72–81. ACM, 2008.
4. R. D. Blumofe, C. F. Joerg, B. C. Kuszmaul, C. E. Leiserson, K. H. Randall, and Y. Zhou, "Cilk: An efficient multithreaded runtime system," Journal of Parallel and Distributed Computing, vol. 37, no. 1, pp. 55–69, Aug. 1996.
5. Q. Chen, M. Guo, Q. Deng, L. Zheng, S. Guo, and Y. Shen. Hat: history-based auto-tuning mapreduce in heterogeneous environments. *The Journal of Supercomputing*, pages 1–17, 2013.
6. Q. Chen, and M. Guo. Adaptive workload-aware task scheduling for single-ISA asymmetric multicore architectures. ACM Transactions on Architecture and Code Optimization, 11(1):8, 2014.
7. M. De Vuyst, R. Kumar, and D. Tullsen, "Exploiting unbalanced thread scheduling for energy and performance on a cmp of smt processors," in *Proceedings of the 2006 IEEE International Parallel and Distributed Processing Symposium*. IEEE, 2006, pp. 10–20.
8. M. Frigo, C. E. Leiserson, and K. H. Randall, "The implementation of the Cilk-5 multithreaded language," in *Proceedings of the ACM SIGPLAN '98 Conference on Programming Language Design and Implementation*. Montreal, Canada: ACM, Jun. 1998, pp. 212–223.
9. Y. Guo, R. Barik, R. Raman, and V. Sarkar, "Work-first and help-first scheduling policies for async-finish task parallelism," in *Proceedings of the 2009 IEEE International Parallel and Distributed Processing Symposium*. IEEE Computer Society, 2009, pp. 1–12.
10. S. Hofmeyr, C. Iancu, and F. Blagojević, "Load balancing on speed," in *Proceedings of the 15th ACM SIGPLAN symposium on Principles and Practice Of Parallel Programming*. ACM, 2010, pp. 147–158.
11. J. A. Joao, M. A. Suleman, O. Mutlu, and Y. N. Patt. Bottleneck identification and scheduling in multithreaded applications. In *Proceedings of the 17th International Conference on*

Architectural Support for Programming Languages and Operating Systems, pages 223–234, 2012.
12. D. Koufaty, D. Reddy, and S. Hahn, "Bias scheduling in heterogeneous multi-core architectures," in *Proceedings of the 5th European conference on Computer systems*. ACM, 2010, pp. 125–138.
13. N. Lakshminarayana, J. Lee, and H. Kim, "Age based scheduling for asymmetric multiprocessors," in *Proceedings of the Conference on High Performance Computing Networking, Storage and Analysis*. ACM, 2009, p. 25.
14. T. Li, D. Baumberger, D. Koufaty, and S. Hahn, "Efficient operating system scheduling for performance-asymmetric multi-core architectures," in *Proceedings of the 2007 ACM/IEEE Conference on SuperComputing*. ACM, 2007, pp. 1–11.
15. J. Liu and C. Liu, Bounds on scheduling algorithms for heterogeneous computing systems. Dept. of Computer Science, University of Illinois at Urbana-Champaign, 1974.
16. M. Mahoney. Data compression programs, 2013.
17. C. Maia, L. Nogueira, and L. M. Pinho. Scheduling parallel real-time tasks using a fixed-priority work-stealing algorithm on multiprocessors. In *the 8th International Symposium on Industrial Embedded Systems*. IEEE, 2013.
18. S. Mattheis, T. Schuele, A. Raabe, T. Henties, and U. Gleim. Work stealing strategies for parallel stream processing in soft real-time systems. In *Architecture of Computing Systems*, pages 172–183. Springer, 2012.
19. A. Navarro, R. Asenjo, S. Tabik, and C. Caşcaval. Load balancing using work-stealing for pipeline parallelism in emerging applications. In *Proceedings of the 23rd international conference on Supercomputing*, pages 517–518. ACM, 2009.
20. A. Rosenberg and R. Chiang, "Toward understanding heterogeneity in computing," in *Proceedings of the 2010 IEEE International Parallel and Distributed Processing Symposium*. IEEE, 2010, pp. 1–10.
21. J. C. Saez, A. Fedorova, M. Prieto, and H. Vegas. Operating system support for mitigating software scalability bottlenecks on asymmetric multicore processors. In *Proceedings of the 7th ACM international conference on Computing frontiers*, pages 31–40. ACM, 2010.
22. J. C. Saez, M. Prieto, A. Fedorova, and S. Blagodurov. A comprehensive scheduler for asymmetric multicore systems. In *Proceedings of the 5th European conference on Computer systems*, pages 139–152. ACM, 2010.
23. D. Shelepov, J. Saez Alcaide, S. Jeffery, A. Fedorova, N. Perez, Z. Huang, S. Blagodurov, and V. Kumar. Hass: a scheduler for heterogeneous multicore systems. ACM SIGOPS Operating Systems Review, 43(2):66–75, 2009.
24. M. Suleman, O. Mutlu, M. Qureshi, and Y. Patt, "Accelerating critical section execution with asymmetric multi-core architectures," in *Proceeding of the 14th international conference on Architectural Support for Programming Languages and Operating Systems*. ACM, 2009, pp. 253–264.
25. K. Van Craeynest, A. Jaleel, L. Eeckhout, P. Narvaez, and J. Emer. Scheduling heterogeneous multi-cores through performance impact estimation (pie). In *Proceedings of the 39th International Symposium on Computer Architecture*, pages 213–224. IEEE Press, 2012.
26. L. Zheng, Y. Lu, M. Ding, Y. Shen, M. Guo, and S. Guo. Architecture-based performance evaluation of genetic algorithms on multi/many-core systems. In *the 14th International Conference on Computational Science and Engineering*, pages 321–334. IEEE, 2011.

Chapter 6
Load Balancing for Heterogeneous Parallel Architecture

Abstract Besides traditional CPU-based parallel computer, heterogeneous parallel architectures that consists of both CPU and GPGPU are used in many emerging large-scale clusters/supercomputers. In order to better utilize both the CPU and GPU, an application could divide and distribute its workload to the two types of hardware at the same time. However, it is not trivial to find an optimal allocation for all the applications offline, because applications often have various characters thus different applications have different speedup ratio on GPGPU compared with that on CPU. In order to solve this problem, this chapter presents the techniques that can balance the application workload across heterogeneous hardware.

6.1 Background and Existing Problems

Heterogeneous parallel architecture that consists of traditional CPU and accelerators are becoming more and more popular in large-scale datacenters/clusters. Accelerators, such as GPGPU and FPGA, often could provide much higher processing power if the target application can adapt to the architecture of the accelerator. Many researchers have.

Although heterogeneous systems with CPU and accelerator are widely used, programmers may not utilize them efficiently since it is challenging for the programmer to split and balance the workload between CPU and the accelerator. Without loss of generality, in this chapter, we use GPGPU as the represent accelerator.

In order to utilize both CPU and GPU efficiently, emerging task schedulers often leverage data parallelism to distribute the workload for heterogenous parallel architectures. Leveraging data parallelism, the task scheduler partitions the data (workload) of an application into multiple parts, and allocates different parts to different devices so that they can process the data concurrently.

Many scheduling strategies, either static or dynamic, have been proposed to balance the workload between CPU and GPU. In a static policy, the workload is split

Part of contents in this chapter has been published through International Workshop on Programming Models and Applications for Multicores and Manycores. Reprinted from Ref. [14], with permission from ACM.

© Springer Nature Singapore Pte Ltd. 2017

Q. Chen and M. Guo, *Task Scheduling for Multi-core and Parallel Architectures*,
https://doi.org/10.1007/978-981-10-6238-4_6

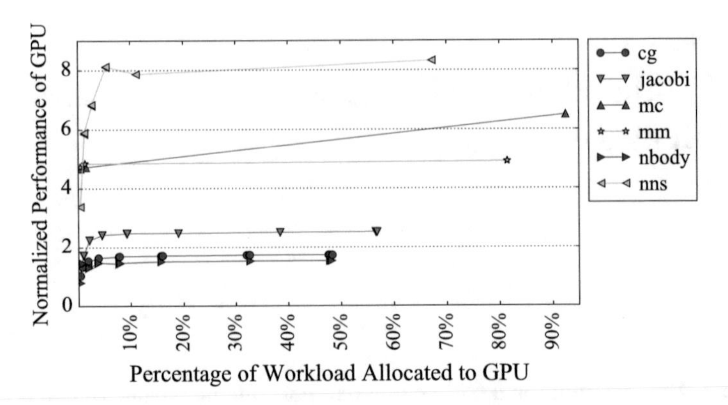

Fig. 6.1 Performance of GPU when it is allocated different amount of workload (the higher the better)

and distributed to CPU and GPU statically. Static strategies often cause unbalanced workload because it is chanllenging to predict the performance of GPU for a particular program without knowing the details of the runtime. GPU is highly sensitive to scale of the workload. The performance is unknown to the scheduler without previous executions.

In order to better illustrate the above problem, as shown in Fig. 6.1, we allocate different percentages of an application's workload to GPU and compare the performance. The detailed experimental platform and the benchmarks can be found in Sect. 6.5. In the figure, for an application, the x-axis indicates the percentage of its workload allocated to GPU and y-axis is the performance of GPU normalized to its performance on CPU. Let IPC_{gpu} and IPC_{cpu} represent the instruction-per-cycle (IPC) of the application on GPU and on CPU, respectively. The normalized performance of GPU (denoted by P_{gpu}) is calculated in Eq. 6.1.

$$P_{gpu} = \frac{IPC_{gpu}}{IPC_{cpu}} \tag{6.1}$$

Observed from the figure, the performance of GPU increases while more workloads are allocated to it. Furthermore, its performance increases rapidly at the beginning of the curve but goes stable at the end of the curve. The poor performance of GPU when its workload is small is due to the synchronization overhead between CPU and GPU. To this end, if the workload of an application is divided into a great amount of small chunks, and then balance the workloads across CPU and GPU in fine granularity, the computational ability of GPU is not fully utilized. It is beneficial if we can find the optimal allocation and assign a large chunk of the whole workload to GPU once instead of splitting it into many small chunks.

However, it is challenging to find the optimal allocation because the performance of GPU for an application varies non-linearly while it is allocated different amount of workload. If the scheduling granularity is too small, the synchronization overhead hurts the overall performance. On the contrary, if the scheduling granularity is too large, it is hard to balance the workload. To conclude, both the *scheduling granularity* and the *load-balancing* significantly impacts the overall performance of an application on heterogeneous parallel architecture. In this chapter, considering the two factors, we introduce a *Heterogeneous-Aware Task Scheduler (HATS)* that balances workload across CPU and GPU while still fully utilizing both of them.

6.2 Prior Solutions

GPU programming is becoming an important issue in the parallel programming area. Some programming language extensions like CUDA [11], Brook+ [3] and OpenCL [10] are published to utilize the hardware's raw performance. But the performance of these extensions are not portable between devices because they are close to the hardware. They can get good performance but programmers need to have a good understanding of the hardware to efficiently use both CPU and GPU.

GPU's performance characteristics have been deeply studied. [12] explored the optimizations of GPU and shows that a fully optimized GPU program is much faster than an unoptimized GPU program. [15] and [6] used performance models and analyzed the instructions that generated by NVCC to predict the performance of GPU statically. [2] made a tool that does the analysis automatically. [7] extended [6] by integrating a power model into their performance model to get more detailed analysis. These models can be used in static analysis of the performance of GPU but it is not useful at runtime because it introduces much overhead.

GPU+CPU co-scheduling is also getting attention with the increasing usage of GPU in high performance computing. Several platforms are designed and implemented to combine the processing power of CPU and GPU. Mars [5] used Mapreduce as its programming paradigm. StarPU [1], Qilin [8] and Scout [9] offered different methods to map tasks to CPU and GPU. OmpSS [4] extended OpenMP to provide co-scheduling ability. These platforms require the programmer to rewrite their code using a new programming language in the case of StarPU or Scout or using specific APIs in Mars and Qilin. In the following of this section, we introduce several widely-used techniques that balance the workload of an application to CPU and GPU.

6.2.1 Static Scheduling

Static workload scheduling, which statically splits and allocates the workload to CPU and GPU before the program execution, is the traditional policy. In this policy, the way the workload is partitioned is determined by either the programmer or the scheduler. Figure 6.2 shows the traditional static workload scheduling for heterogeneous parallel architecture.

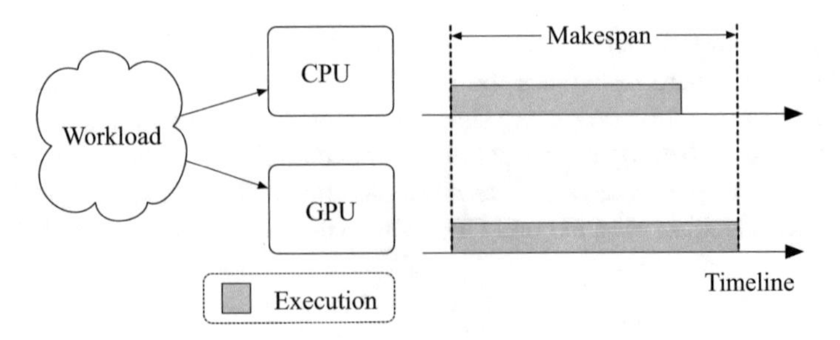

Fig. 6.2 Static scheduling policy for heterogeneous parallel architecture

Observed from the figure, we can find that static scheduling only considers the factor of *scheduling granularity* but often fails on balancing the workload. CPU and GPU do not complete their tasks at the same time. This policy does not work well for heterogeneous architecture because the performance of GPU varies for different algorithms, different workload sizes and different application implementations. It is not possible to find an optimal static allocating that performs the best for all the applications.

Besides the naive static scheduling policy, researchers have also proposed several variations to improve its performance. For instance, a scheduler could record the execution time of an application on CPU and GPU in previous executions, and calculate the performance of CPU and GPU for the application offline [8]. However, this method is not helpful if a brand-new application is executed. In addition, even for the same application, the optimal workload allocation may vary with the different inputs. In another static scheduling variation, the scheduler analyzes the code generated by the compiler to calculate the performance of a specific program on a specific GPU. However these methods will fail if the program is executed on other hardware environments because the previous execution time and the offline analysis are no longer accurate.

To conclude, the static scheduling policy does not need synchronization between CPU and GPU, thus has small overhead. However, the potential unbalanced workload may degrade the overall performance.

6.2.2 *Quick Scheduling*

In order to balance the workload, dynamic scheduling policy that decides the workload allocation based on the performance of CPU and GPU at runtime is proposed. Quick scheduling policy [13] is a popular dynamic scheduling policy for heterogeneous architecture.

Figure 6.3 shows the processing flow of an application on heterogeneous architecture with the quick scheduling policy. As shown in the figure, with the quick scheduling policy, the scheduler first allocates a small portion of the workload to

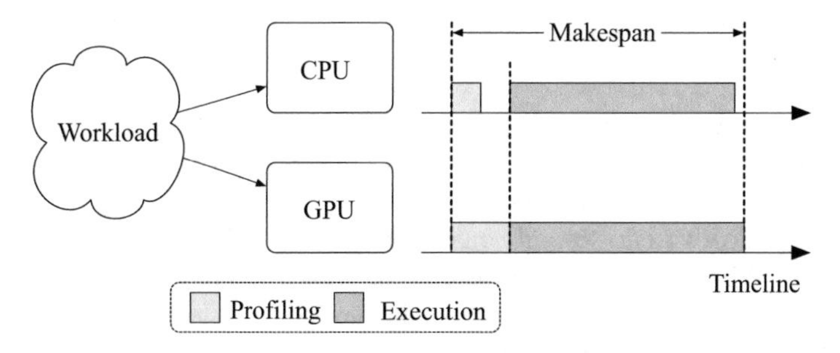

Fig. 6.3 Quick scheduling policy for heterogeneous parallel architecture

CPU and GPU to profile their performance for the given application. Based on the collected performance information, the remaining of the workload is split and allocated to them.

Let IPC_{cpu} and IPC_{cpu} represent the instruction-per-cycle (IPC) of the application on CPU and GPU in the profiling step respectively, and let W represent the remaining workload of the application. Equation 6.2 calculates the amount of workloads that should be allocated to CPU and GPU, denoted by W_{cpu} and W_{gpu}, in the execution step.

$$W_{cpu} = W \times \frac{IPC_{cpu}}{IPC_{cpu} + IPC_{gpu}}, W_{gpu} = W \times \frac{IPC_{gpu}}{IPC_{cpu} + IPC_{gpu}} \tag{6.2}$$

Observed from the figure, quick scheduling policy tries to maximize the workload assigned to GPU while achieving good load-balance. Although it adjusts the partition according to runtime information, the performance of GPU and GPU it calculates may not be accurate. Because the performance of GPU changes when the workload changes, GPU may perform differently in the profiling step and in the execution step.

Recall that the performance of GPU becomes stable while its workload increases (Sect. 6.1). Therefore, in order to improve the accuracy of performance prediction for GPU in the execution step, we can increase the amount of the workload used in the profiling step. However, it is challenging to find the optimal amount of workload to use in the profiling step. If too much workload are used in the profiling step, the overall performance could be seriously damaged because the partition in the first step is usually not balanced. On the contrary, if too few workload are used in the profiling step, the obtained performance of GPU is not accurate and thus results in the poor load balance in the execution step.

To conclude, the quick scheduling policy does not introduce much synchronization overhead. Compared with static scheduling policy, it can better balance the workload across CPU and GPU, thus it is able to improve the performance of parallel applications. However, because the profiled performance of GPU in the profiling step may not accurate, the workload is not perfectly balanced across CPU and GPU in the execution step.

6.2.3 Split Scheduling

In order to achieve the optimal load balance, researchers proposed split scheduling policy [13]. Figure 6.4 shows the split scheduling policy for heterogeneous parallel architecture. As shown in the figure, in the policy, the whole workload of an application is divided into several equal-sized chunks and the chunks are processed sequentially. When the scheduler allocates the workload in a chunk to the CPU and GPU, the percentage of the workload allocated to GPU is calculated according to the GPU performance predicted from the execution of the previous chunk. The amount of workload allocated to CPU and GPU in a chunk can be calculated in the same way in Eq. 6.2.

Observed from the figure, we can find that the split scheduling policy tries to get the accurate performance of GPU over CPU. Compared with the static scheduling policy and the quick scheduling policy, split scheduling can best balance the workload across the CPU and GPU, because it profiles in each chunk and collect much more information than the other two polices. However, because it splits the whole overhead into many chunks, CPU and GPU need to synchronize with each other at the end of the processing of each chunk. In this case, the split scheduling policy introduces heavy synchronization overhead.

It is easy to find that the performance of the split scheduling is sensitive the the size of the workload chunks. If the chunk size is too small, CPU and GPU needs to synchronize with each other frequently and the overhead may degrade the overall performance. On the other hand, if the chunk size of too large, the application may complete before the best partitioning is achieved. In this case, the performance of the application is also sub-optimal. In real-system scenario, it is challenging to find the optimal chunk size because different applications often have different features thus have different optimal chunk sizes.

To conclude, split scheduling policy only considers the factor of *load balancing* and fails to consider to *scheduling granularity*. Therefore, split scheduling policy is not able to achieve the best performance for applications on CPU+GPU heterogeneous architecture.

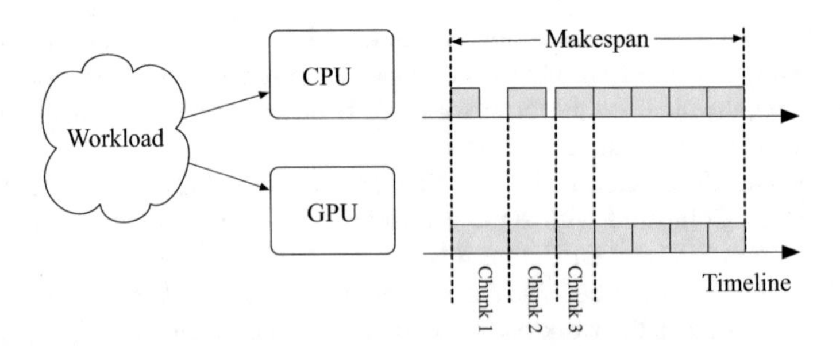

Fig. 6.4 Split scheduling policy for heterogeneous parallel architecture

6.2.4 *FinePar*

Besides the above techniques, Zhang et. al [16] proposed *FinePar* that partitions irregular workloads between CPU and GPU through irregularity-aware performance modeling and online-tuning. In this way, FinePar is able to achieve both device-level and thread-level load balance. In FinePar, Zhang et. al [16] designed a program transformation to automatically transform the given OpenCL program to enable fine-grained partitioning; built performance models to predict the performance of the CPU and GPU given any specific fine-grained partitioning; and designed an auto-tuner to guide the fine-grained workload partitioning for load balancing between the CPU and GPU.

It is worth noting that FinePar only works for matrix-based applications. In the following of this section, I will introduce the performance modeling and auto-tuning techniques in FinePar. The program transformation is not closely related to this work. If you are interested, you can refer to the original paper of FinePar [16].

6.2.4.1 Performance Modeling

FinePar uses linear regression to build performance models for its low overhead. In the performance model, features that are closely related with the OpenCL programming model and those that represent irregularity of the workload are selected. More specifically, four features are selected:

1. The average workload for a work-item (*AW*). A work-item in OpenCL is similar to a *thread* in CUDA.
2. The variance of the distribution of non-zero elements across the rows (*VW*).
3. The number of work-items in the computation domain (*NW*).
4. The size of the whole workload (*SW*).

The targeted sparse matrix applications have irregular memory access pattern, which affects cache performance and the main memory bandwidth utilization. However, the memory access pattern is not captured by the linear regression model. Despite its importance, the memory access pattern depends on the distribution of the non-zero elements and the interleaved execution of the threads, which is expensive to profile and hard to model. Hence, to circumvent this problem, the training matrices are categorized into quasi-diagonal matrices and non-quasi-diagonal ones, which are referred as Type 1 and Type 2 matrices, respectively. FinePar builds different performance models for each type.

FinePar quantifies the closeness of the non-zero elements to the diagonal in the following way. For each row, FinePar counts the number of non-zero elements whose column is no more than one eighth of the width of the matrix away from the diagonal. FinePar divides the total number of non-zero elements in the matrix by the sum of such numbers for all rows. If the result is larger than the threshold T_{diag}, the matrix is categorized as a Type 1 matrix. Otherwise, it is a Type 2 matrix.

For each type of matrices, FinePar builds a linear regression model for the CPU and one for the GPU. Given a training matrix or graph, a value is chosen for T_f (the partitioning threshold) from $\{16, 32, 64, 128, 256, 512, 1024, 2048\}$ and the matrix can be partitioned into CPU and GPU workloads. The partitioned workloads on the CPU and GPU are then executed to collect execution times for the training, which capture performance degradation due to co-running. Equations 6.3 and 6.4 show the performance models for the GPU and CPU, respectively. The C_i's ($i = 1,, 5$) are the parameters of the model FinePar trains. The graph generator from Graph 500 is used to generate the training data.

$$P_g = C1_g \times AW_g + C2_g \times VW_g + C3_g \times log(NW_g) + C4_g \times log(SW_g) + C5_g \quad (6.3)$$

$$P_c = C1_c \times AW_c + C2_c \times VW_c + C3_c \times log(NW_c) + C4_c \times log(SW_c) + C5_c \quad (6.4)$$

6.2.4.2 Applying FinePar Online

Given the input data, the goal of online tuning is to select the threshold for fine-grained partitioning to achieve the best performance. It consists of two stages: (1) matrix category detection, and (2) threshold search. The detection stage determines the matrix category and subsequently the performance models to use. The search stage leverages the performance models to predict performance given a threshold and search for the optimal threshold.

When determining the category of the input matrix online, in order to minimize the overhead, FinePar samples a number of rows from the input matrix and only counts the non-zero elements close to the diagonal for the sampled rows. For the quantification to determine the category, FinePar scales down the total number of non-zero elements according to the sampling ratio.

Threshold search uses the hill climbing algorithm to search for the optimal threshold. FinePar first chooses an initial value for T_f such that the ratio between the numbers of non-zero elements in the two partitioned workloads matches the ratio of the peak performance between the CPU and GPU. It then uses the performance model to estimate the execution time given T_f , $(T_f - step)$, and $(T_f + step)$ as the threshold, respectively. If T_f produces the optimal performance, the tuning process terminates. Otherwise, T_f is assigned one of the two other values, which yields better performance.

According to the above description, we can find that FinePar requires offline training to achieve good partitioning between CPU and GPU. Furthermore, it is not applicable for applications that do not use matrix as input. In this following of this chapter, we introduce a pure online heterogeneous-aware task scheduling policy that balance the workload across CPU and GPU without any offline training.

6.3 Heterogeneous-Aware Task Scheduling

In order to balance the workload while minimizing the number of synchronization between CPU and GPU, we present the *Heterogenoues-Aware Task Scheduling* (HATS) policy in this section.

Figure 6.5 shows the heterogeneous-aware task scheduling policy, HATS, for heterogeneous architecture. HATS breaks the whole execution into several steps. In the first step, it executes a small portion of the workload with static partition and collects the execution time like quick scheduling. Instead of partitioning the remaining workload with the ratio it calculates in the first step, HATS executes the next step whose size of workload is doubled. In this method, HATS profiles the performance of GPU for different sizes of workload. HATS continues profiling and doubles the size of workload in each step. To find the stable point of performance, HATS calculates the variance of the current and the previous performance ratio in each step. If the variance is smaller than the threshold, or the remaining workload is smaller than the workload HATS tries to profile, HATS will stop profiling and execute the remaining workload. Algorithm 10 shows the policy in peusocode.

Algorithm 10 Heterogeneous-Aware Task Scheduling (HATS)

1: **if** This is the first step **then**
2: Take a small portion of the workload and partition with static partition.
3: Record the size of the workload as s
4: **else if** There is remaining workload **then**
5: Calculate the performance ratio of the previous execution.
6: Calculate the partition of the current execution according to the performance ratio.
7: Calculate the variance of the two partitions.
8: **if** The variance is small enough or $s*2$ is larger than $1/2$ of remaining workload **then**
9: Partition the remaining workload
10: **else**
11: $s \leftarrow s*2$
12: Partition s
13: **end if**
14: **end if**

Suppose HATS schedules a program that has 65536 iterations which can be executed in parallel on a CPU+GPU heterogeneous system. In the first step, HATS takes $1/128$ (this parameter is set statically before execution) of the iterations, which is 512. Then HATS splits these iterations with static partition and allocates them to CPU and GPU. We assume that the partition ratio is 1:1 in this case. In another word, HATS assigns 256 iterations to the CPU and 256 iterations to the GPU. HATS also transfers the required data of these iterations to the GPU before launching GPU kernel. After the CPU and the GPU finish their work, HATS synchronizes, transfers the result from the GPU to the CPU and collects the execution times of both devices. Then it calculates the iterations of each device completes per unit time (second, for example).

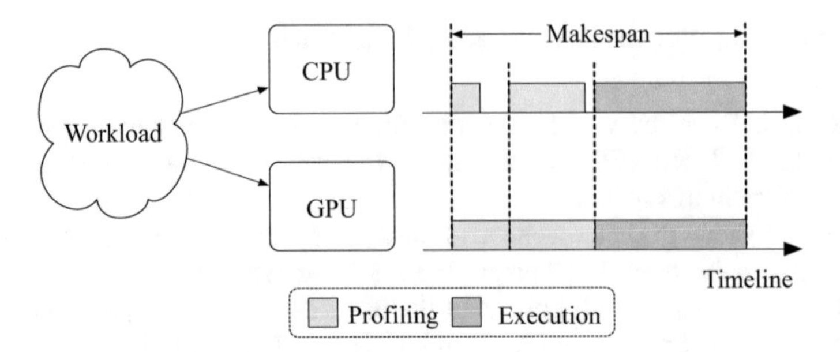

Fig. 6.5 Design of HATS for heterogeneous parallel architecture

In the second step, HATS compares the partition it calculated with the partition it used in the previous step. It calculates the variance of the two partitions. If the variance is smaller than the threshold, the performance of GPU is stable enough for HATS to make a good partition. Then It partitions the remaining workload (65024 iterations) according to it. If not, HATS will use the performance ratio it calculated in the previous step to partition 1024 ($2 * 512$) iterations.

HATS keeps profiling until the next step will take more than $1/2$ of remaining iterations. If the next step takes more than $1/2$ of remaining iterations, HATS will simply execute the remaining workload with current partition because it may not be possible for HATS to find a stable partition without degrading the performance of GPU. HATS tries to maximize the performance of GPU by assigning a large amount of workload to GPU.

HATS profiles the performance of GPU in an asymptotic way. HATS tries to find the stable point of GPU's performance curve by assigning different sizes of workload to GPU. In quick scheduling, scheduler assumes that the performance of GPU is a constant but it is not. HATS keeps profiling until the performance of GPU is stable to estimate the performance. In every step, HATS adjusts the partition to get closer to the best partition. When it finds the stable point, it stops profiling and partitions the remaining workload according to the best partition it can get.

6.4 Comparison of the Scheduling Policies

In previous sections, we introduced four existing task scheduling policies for heterogeneous architecture: static scheduling policy, quick scheduling policy, split scheduling policy, and HATS. In this section, we systematically compare the four policies from three aspects: *initial partition*, *performance prediction*, *partition method*, *load balancing*, and *synchronization*.

Initial Partition. All the four introduced policies adopt static partition in the first step, since the scheduler has no knowledge about the performance of GPU and CPU when a brand-new application is submitted. The static scheduling policy partitions the whole workload into two parts and allocates them to CPU and GPU directly. On the other hand, the dynamic scheduling policies (i.e., quick scheduling policy, split scheduling policy and HATS) only partition a small subset of the whole workload in the first step to avoid performance loss due to the unbalanced workload.

Performance prediction. Static scheduling policy uses the static partition all the time. Dynamic scheduling policies (Quick scheduling policy, split scheduling policy and HATS) use the obtained performance of CPU and GPU from the profiling step to partition future unexecuted workload. During the partitioning, static scheduling policy and quick scheduling policy assume that the performance of GPU is consistent, which is not true according to our experiment in Sect. 6.1. On the contrary, split scheduling policy and HATS do not have this assumption and calculates the actual IPC on the CPU and GPU to calculate their performance. To this end, they can better partition and allocate the workload, thus often perform better than static scheduling policy and quick scheduling policy.

Partition method. Quick scheduling policy splits the whole workload into a small chunk and a large chunk. It profiles the small chunk and executes the big chunk with the performance of CPU and GPU collected from the execution of the small chunk. In the split scheduling policy, the execution of an application is divided into a given number of sequential steps and the size of the workload executed in each step is the same. HATS dynamically decides the number of steps. HATS calculates the variance of partitions to check whether the performance of GPU is stable. If it is stable, HATS stops profiling and directly allocates the remaining workload to the CPU and GPU.

Load Balancing. In term of load balancing, static scheduling policy often performs the worst, because the workload is allocated without considering the actual performance of CPU and GPU. Meanwhile, quick scheduling policy is not able to optimally balance the workload across CPU and GPU. This is mainly because quick scheduling policy predicts the performance of CPU and GPU according to their performance with small workload. As shown in Fig. 6.1, when the workload is small, the performance of GPU changes seriously. It is not accurate to predict the performance of GPU with large workload using its performance with small workload. To this end, quick scheduling policy often suffers from sub-optimal load balancing. On the contrary, HATS increases the size of the workload in each step exponentially to find the stable point of the curve and adjusts the partition according to the execution time. Once the partition is stable, HATS can safely execute the remaining workload and get good load-balance. Similarly, split scheduling policy can also balance the workload.

Synchronization. In static scheduling policy, CPU and GPU only need to synchronize with each other for one time. In quick scheduling policy, CPU and GPU synchronize with each other for two times (one after the profiling step, and the other one after the execution step). In split scheduling policy, the number of synchronization is determined by the number of chunks the overall workload is divided. If the overall workload is divided into m chunks, CPU and GPU synchronize with each other for m times (one time at the end of each chunk). Compared with the split scheduling policy, HATS requires CPU and GPU to synchronize with each other for only a few times because the size of the workload in each step is increased exponentially. The number of synchronizations in split scheduling policy is linear to the smallest workload while the number of synchronization with HATS is only logarithmic to the smallest workload. HATS does not degrade the performance of GPU much because HATS increases the size of the workload assigned to GPU in each step. It uses a fixed partition only after the performance of GPU goes stable.

6.5 Performance of Dynamic Scheduling Policies

In this section, we compare the performance of the introduced dynamic task scheduling policies for heterogeneous architecture. We omit the performance of static task scheduling policy, because it is hard to find the appropriate allocation for it.

6.5.1 Experimental Setup

The detailed experiment setups are summarized in Table 6.1. In our experiment, we use six widely-used benchmarks, which are listed in Table 6.2, to evaluate the performance of quick scheduling policy, split scheduling policy and HATS. In the 6 benchmarks, *cg*, *jacobi* and *mm* are classic matrix algorithms. *nbody* is a classic physics problem that simulates a dynamical system of particles, usually under the influence of physical forces, such as gravity. *mc* is the Monte Carlo method for European option pricing and *nns* is a search algorithm widely used in machine learning. We choose these benchmarks because they are widely used in emerging scientific applications that demand high computational ability. Heterogenous architecture that consists of both CPU and GPU can provide such ability.

In order to evaluate the three scheduling policies, for each benchmark, we implement three versions: single-thread CPU-only version, GPU-only version, and heterogeneous version. The CPU-only version is implemented with C language; the GPU-only version is implemented with CUDA. In the heterogeneous version, we implement the quick scheduling policy, split scheduling policy and HATS to manage the scheduling. In our implementation, the initial workload for quick scheduling

Table 6.1 Hardware and software specifications

	Specifications
CPU	Intel Xeon E5620 @ 2.4GHz
GPU	Nvidia Tesla M2090 @ 1.3GHz
CPU code compiler	GCC 4.6.3
GPU code compiler	NVCC 5.0
Operating System	Debian Wheezy (Linux 3.2)

Table 6.2 Benchmarks used to evaluate the performance of the scheduling policies.

Benchmark	Description	Workload
cg	Conjugate Gradient method	16Kx16K matrix
jacobi	Jacobi method	16Kx16K matrix
mc	European Option Pricing	64M iterations
mm	Matrix Multiplication	Two 1Kx1K matrix
nbody	N-Body Simulation	16K bodies
nns	Nearest Neighbour Search	16K points with 16 K queries

policy and HATS is 1/128 of the whole workload; for split scheduling policy each chunk has 1024 iterations. These numbers are determined according to the paper that proposed the techniques.

It is worth noting that, for an application, we report its performance to be the time used by its accelerated parallel region, including kernel launching overhead, data transferring time, scheduling overhead and computation time. The time used for initialization and data preparation is excluded from the execution time, because they are not affected by the scheduling policy. For a benchmark, we use relative time (speedup) instead of absolute time (seconds) in our results and the speedup is the performance normalized with the performance of CPU-only version.

6.5.2 *Performance*

Figure 6.6 shows the performance of quick scheduling policy, split scheduling policy and HATS for all the benchmarks. Observed from the figure, we can find that HATS outperforms the quick scheduling policy and split scheduling policy for all the benchmarks. HATS improves the performance of benchmarks ranging from 1.1 to 86.7% (33.4% on average) compared with quick scheduling policy, and ranging from

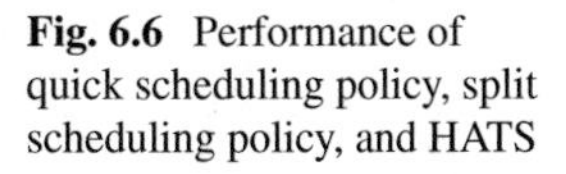
Fig. 6.6 Performance of quick scheduling policy, split scheduling policy, and HATS

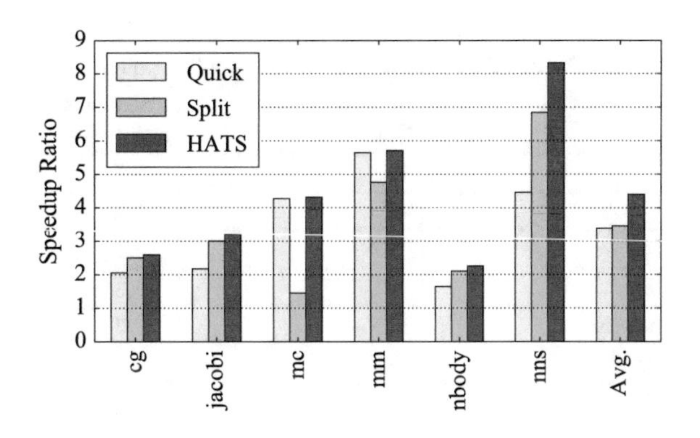

3.6 to 197.3% (42.7% on average) compared with split scheduling policy. In more detail, Table 6.3 summarizes the performance improvement of HATS compared with quick scheduling and split scheduling.

Besides the improvement over previous scheduling policies, Table 6.3 also gives the percentage of workload used in the profiling step to search the optimal workload allocation in HATS. HATS adjusts the percentage of workload used in the profiling step accordingly. Observed from the table, for *cg*, *jacobi*, *nbody* and *nns*, HATS profiles $1/4$ to $1/3$ of the workload but it only profiles $1/50$ of *mm*'s workload. HATS only profiles when needed rather than profiling the 100% of workload like split scheduling.

In addition, observed from the table, we can find that HATS performs better when the profiling step is short. For instance, when the profiling step only processes 3.1% of the whole workload, HATS improves the performance of the benchmarks up to 197.3% compared with the split scheduling policy. This is because the number of synchronizations is small in HATS when the profiling step is short. On the contrary, the number of synchronizations is always large in Split scheduling policy.

From Table 6.3 we can also observe that HATS only improve the performance of *mc* and *mm* by 1.1%. After looking into the workload of *mc* and *mm*, we find that the workload of *mc* and *mm* is pretty large, and 1/128 of the whole workload used in the profiling step is already large enough to fully utilize the GPU. In this case, the performance of GPU in the execution step can be precisely predicted with the performance information obtained in the profiling step. Therefore, Quick scheduling policy can efficiently balance the workload across CPU and GPU for *mc* and *mm*. Meanwhile, HATS still performs slightly better than the Quick scheduling policy because it predicts the performance of GPU more precisely with an extra profiling step.

Compared with Quick scheduling policy, Split scheduling policy can better balance the workload. However, the frequent synchronizations and the small workload chunks degrade the performance of GPU. For instance, *mc* has 64M iterations and each workload chunk only has 1024 iterations. In this case, split scheduling policy splits the overall workload into $\frac{64M}{1024} = 64K$ small chunks, and executes these chunks sequentially. The large number of chunks results in the large number of CPU-GPU

Table 6.3 Improvement over previous scheduling policies

Benchmark	Improvement to Quick(%)	Improvement to Split(%)	Percent of Profiling(%)
cg	26.1	3.6	37.9
jacobi	47.6	6.9	26.8
mc	1.1	197.3	3.1
mm	1.1	19.8	3.1
nbody	37.5	7.2	26.8
nns	86.7	21.6	25.0

synchronization, which in turn significantly degrades the performance of the benchmarks. For other benchmarks that have small workload (e.g., nbody), split scheduling policy performs much better.

6.5.3 *Effectiveness of Balancing Workload*

In this subsection, we evaluate the effectiveness of the three dynamic scheduling policies in balancing workload across CPU and GPU. To measure the effectiveness in balancing workload, we calculate the *imbalance degree* of an allocation, denoted by D, in Eq. 6.5. In the equation, T_{cpu} and T_{gpu} is the execution time of the allocated workload on CPU and GPU respectively. Obviously, the smaller the imbalance degree is, the better the workload is balanced across CPU and GPU in heterogeneous architecture.

$$D = \frac{|T_{cpu} - T_{gpu}|}{\max\{T_{cpu}, T_{gpu}\}} \tag{6.5}$$

Fig. 6.7 shows the imbalance degrees of all the benchmarks when they are scheduled with Quick scheduling policy, Split scheduling policy, and HATS. Observed from the figure, Quick scheduling policy always results in the large imbalance degree. This is mainly because Quick scheduling policy uses the performance of GPU with small workload to predict its performance with large workload. Since the performance of GPU is not predicted precisely in Quick scheduling policy, the workload cannot be perfectly balanced across CPU and GPU. On the contrary, Split scheduling policy and HATS results in much smaller imbalance degree because they can better predict the performance of GPU and thus can better balance the workload.

For Quick scheduling policy, the average imbalance degree is about 1/3. This means that the fast device wastes $1/3$ of the execution time on waiting for the straggler device to complete its workload. This result verifies our argument that a small portion of workload is not enough to estimate the performance of GPU with different workload.

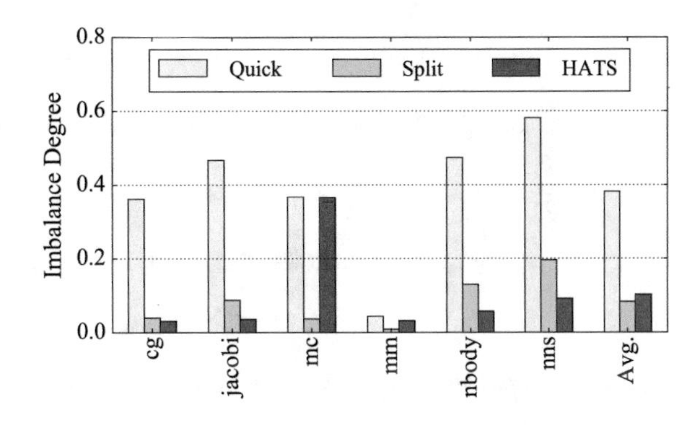

Fig. 6.7 Imbalance degree of the benchmarks when they are scheduled with Quick scheduling policy, Split scheduling policy, and HATS

6.5.4 Effectiveness of Predicting the Performance of GPU

Precisely prediction the performance of GPU is essential for optimally balance the workload across CPU and GPU in heterogeneous architecture. In this subsection, we show the effectiveness of predicting the performance of GPU in HATS.

Figure 6.8 shows the partition variances of HATS in different profiling steps. Partition variance is the variance of the partition in one profiling step and the partition in the previous profiling step. The variance shows the changes of performance of GPU. If the change is small, the variance will be small. Otherwise the variance will be large. HATS tries to find the stable point by keeping profiling and calculating the variance.

The percentage of profiling is related to the threshold. If the threshold is high, the percentage will be lower, but the load-balance will be worse. If the threshold is low, the percentage will be higher. We set the threshold to $5 * 10^{-5}$ but other values can be used to find a balance point that keeps load-balance with only a few synchronizations.

If the variance is smaller than the threshold, HATS will stop profiling. The figure shows that the first profiling step for mm is quite accurate that HATS only needs one more profiling step to ensure the partition is correct. Others converge to the threshold quickly and five steps is usually enough for good partition.

6.5.5 Impact of Profiling Granularity

For an application, HATS first profiles it using small workload and increases the size of workload used in a profiling step exponentially. The start point may affect the performance of HATS. If HATS starts to profile an application with a small workload, it avoids the imbalance in the first step. However, starting from small workload means that HATS requires more steps to reach the stable point. It is not trivial to determine whether HATS should starts to profile the application with a small workload. To this

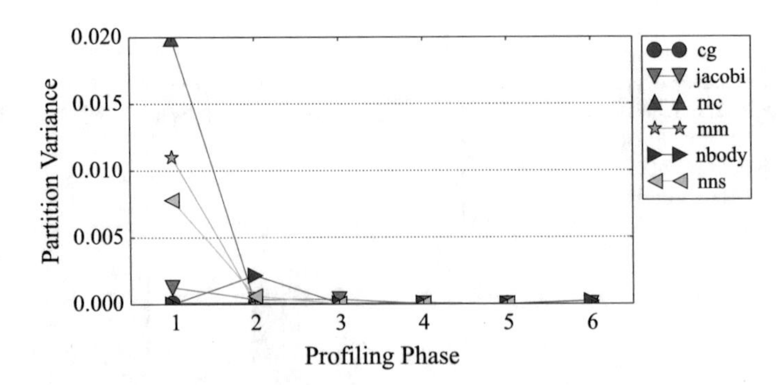

Fig. 6.8 Partition variance of HATS in each profiling step (the smaller the better)

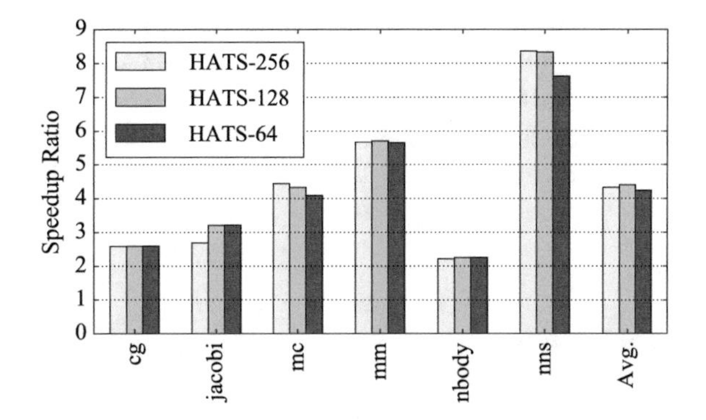

Fig. 6.9 Comparison of different setting of initial workload

end, in this subsection, we evaluate the performance of HATS with three different start points: 1/256, 1/128 and 1/64 of the total workload. For ease of description, they are referred to be HATS-256, HATS-128 and HATS-64 respectively.

Figure 6.9 shows the performance of all the benchmarks scheduled by HATS-256, HATS-128, and HATS-64. Observed from the figure, the three schedulers performs similar. HATS adapts to different programs without tuning the parameters.

6.6 Summary

Heterogeneous systems with CPU and GPU are becoming popular. It is beneficial to use all the processors to solve a single task by taking advantages of data-parallelism. In this chapter, we introduce four widely-used scheduling policies: static scheduling, quick scheduling, split scheduling, and HATS. Static scheduling policy is not able to fully balance the workload across CPU and GPU. Quick scheduling partially improves the load balance, but still suffers from load imbalance. Split scheduling can efficiently balance the workload, but it incurs severe synchronization overhead.

To solve the problem in existing scheduling policies, in this chapter, we propose HATS, a novel scheduling policy that efficiently balance the workload across CPU and GPU while incurs slight synchronization overhead. Our evaluation using popular scientific benchmarks shows that HATS achieves up to 42.7% performance improvement on average compared with the state-of-the-art co-scheduling policy.

6.6.1 Chapter Highlights

The following highlights of this chapter could be of your interest:

- We analyze the performance characteristics of GPU and current dynamic co-scheduling strategies.

- We propose a dynamic scheduling policy based on profiling for splitting and distributing workload across CPU and GPU with only a few synchronizations between CPU and GPU.
- Evaluation results show that HATS achieves up to 42.7% performance improvement on average compared to the state-of-the-art dynamic scheduling policies for heterogeneous architecture.

References

1. C. Augonnet, S. Thibault, R. Namyst, P. Wacrenier, StarPU: A unified platform for task scheduling on heterogeneous multicore architectures, Concurrency and Computation: Practice and Experience 23 (2) (2011) 187–198.
2. S. S. Baghsorkhi, M. Delahaye, S. J. Patel, W. D. Gropp, W.-m. W. Hwu, An adaptive performance modeling tool for GPU architectures, in: Proceedings of the 15th ACM SIGPLAN Symposium on Principles and Practice of Parallel Programming, PPoPP '10, ACM, New York, NY, USA, 2010, pp. 105–114.
3. I. Buck, T. Foley, D. Horn, J. Sugerman, K. Fatahalian, M. Houston, P. Hanrahan, Brook for GPUs: stream computing on graphics hardware, in: ACM SIGGRAPH 2004 Papers, SIGGRAPH '04, ACM, New York, NY, USA, 2004, pp. 777–786.
4. J. Bueno, L. Martinell, A. Duran, M. Farreras, X. Martorell, R. Badia, E. Ayguade, J. Labarta, Productive cluster programming with OmpSS, Euro-Par 2011 Parallel Processing (2011) 555–566.
5. B. He, W. Fang, Q. Luo, N. K. Govindaraju, T. Wang, Mars: a mapreduce framework on graphics processors, in: Proceedings of the 17th international conference on Parallel architectures and compilation techniques, PACT '08, ACM, New York, NY, USA, 2008, pp. 260–269.
6. S. Hong, H. Kim, An analytical model for a GPU architecture with memory-level and thread-level parallelism awareness, in: Proceedings of the 36th annual international symposium on Computer architecture, ISCA '09, ACM, New York, NY, USA, 2009, pp. 152–163.
7. S. Hong, H. Kim, An integrated GPU power and performance model, in: Proceedings of the 37th annual international symposium on Computer architecture, ISCA '10, ACM, New York, NY, USA, 2010, pp. 280–289.
8. C.-K. Luk, S. Hong, and H. Kim. Qilin: exploiting parallelism on heterogeneous multiprocessors with adaptive mapping. In *Proceedings of the 42nd Annual IEEE/ACM International Symposium on Microarchitecture*, pages 45–55. ACM, 2009.
9. P. McCormick, J. Inman, J. Ahrens, J. Mohd-Yusof, G. Roth, S. Cummins, Scout: a data-parallel programming language for graphics processors, Parallel Computing 33 (10–11) (2007) 648–662.
10. A. Munshi, The OpenCL specification version: 1.2 (2011).
11. C. Nvidia, CUDA C programming guide 5.0 (2012).
12. S. Ryoo, C. I. Rodrigues, S. S. Baghsorkhi, S. S. Stone, D. B. Kirk, W.-m. W. Hwu, Optimization principles and application performance evaluation of a multithreaded GPU using CUDA, in: Proceedings of the 13th ACM SIGPLAN Symposium on Principles and practice of parallel programming, PPoPP '08, ACM, New York, NY, USA, 2008, pp. 73–82.
13. T. R. Scogland, B. Rountree, W.-c. Feng, and B. R. De Supinski. Heterogeneous task scheduling for accelerated openmp. In *Parallel & Distributed Processing Symposium (IPDPS), 2012 IEEE 26th International*, pages 144–155. IEEE, 2012.
14. Z. Wang, L. Zheng, Q. Chen, and M. Guo. CAP: co-scheduling based on asymptotic profiling in CPU+ GPU hybrid systems. *Proceedings of the 2013 International Workshop on Programming Models and Applications for Multicores and Manycores*, pages 107–114. ACM, 2013.

15. Y. Zhang, J. Owens, A quantitative performance analysis model for GPU architectures, in: High Performance Computer Architecture (HPCA), 2011 IEEE 17th International Symposium on, 2011, pp. 382 –393.
16. F. Zhang, B. Wu, J. Zhai, B. He, and W. Chen. Finepar: irregularity-aware fine-grained workload partitioning on integrated architectures. In *Proceedings of the 2017 International Symposium on Code Generation and Optimization*, pages 27–38. IEEE Press, 2017.

Chapter 7
MapReduce for Cloud Computing

Abstract Cloud computing and Big data have attracted serious attention from both researchers and public users. For Cloud computing and Big data, MapReduce is one of the most widely-used scheduling model that automatically divides a job into a large amount of fine-grain tasks, distributes the tasks to the computational servers, and aggregates the partial results from all the tasks to be the final results. It naturally fits the requirement of processing a large amount of data in parallel. However, the performance of MapReduce is often seriously damaged by several straggler tasks that run far slower than other tasks in heterogeneous environments where the servers have different computational ability. To this end, in this chapter, we discuss the ways to improve the performance of MapReduce in heterogeneous environments. Specifically, we propose a Self-Adaptive MapReduce (SAMR) scheduling policy that can precisely identify the straggler tasks and boot their execution. Experiments on a real-system heterogeneous cluster prove that the proposed technique can significantly improve the performance of MapReduce applications without any program modification.

7.1 Introduction to MapReduce

In Cloud computing and Big data era, in order to provide satisfactory service, many applications need to process a high volume of data. In order to complete the data processing in the acceptable time, users prefer to use a large amount of computers concurrently. This need has promoted the development of MapReduce, which is one of the most popular programming and scheduling model to process and generate large data sets [17]. MapReduce enables users to specify a map function that processes a key/value pair to generate a set of intermediate key/value pairs, and a reduce function that merges all the intermediate values associated with the same intermediate key

---

Part of contents in this chapter has been published through The Journal of Supercomputing. Reprinted from Ref. [9], with permission from Springer. Figures 7.1 and 7.4 in this chapter have been published through The Journal of Supercomputing. Reprinted from Ref. [9], with permission from Springer.

© Springer Nature Singapore Pte Ltd. 2017

Q. Chen and M. Guo, *Task Scheduling for Multi-core and Parallel Architectures*,
https://doi.org/10.1007/978-981-10-6238-4_7

[17]. MapReduce is used in Cloud Computing in the beginning [3, 5, 12, 29, 30]. It is initiated by Google, together with GFS [25] and BigTable [6] comprising backbone of Google's Cloud Computing platform. Apart from the Cloud Computing platform, MapReduce is also ported to work on GPU and multiprocessors. In addition, it is also extended to solve more loose-coupling problems [10, 11, 13, 14, 26, 28, 32].

For a MapReduce job (i.e., an application that is implemented based on the MapReduce programming model), its data set is divided into many small data sets. When a MapReduce system starts to execute a MapReduce job, the MapReduce scheduler[1] in the system launches a map task for each of the small data sets, and launches a group of reduce tasks to collect the results of all the map tasks. After the division, the MapReduce scheduler distributes these tasks onto different nodes according to the location of the tasks' data sets. In this way, all the nodes (called as *workers*) execute the tasks which are assigned to them in parallel. Since every node needs to execute many tasks, MapReduce scheduler launches a *task scheduler* for each node to manage tasks.

7.1.1 Scheduling Policy in MapReduce

The task scheduler can adopt different policies to schedule the tasks. Apache Hadoop is the most popular programming environment that implements MapReduce. It has implemented multiple scheduling policies with MapReduce. When multiple jobs are submitted to Hadoop, by default, Hadoop adopts first-in-first-out (FIFO) policy to execute the jobs.

FIFO scheduling [4]—In the default FIFO scheduling, all the tasks (including map tasks and reduce tasks) are submitted to the task slots when they are ready and the task slots become free. The task slots run the tasks in a first-come-first-serve manner. One weakness of this scheduling policy is that it is not able to schedule jobs that have different priorities and is not able to guarantee fair sharing of resources between concurrent jobs.

Fair scheduling [32]—In order to ensure fair sharing of resources, Facebook proposed fair scheduling policy. With fair scheduling, each user is assigned a given amount of cluster capacity over a time. Users are able to assign their jobs to different job pools, while each pool is allocated a guaranteed minimum number of Map and Reduce task slots. To achieve this purpose, the fair scheduler provides preemptive technique, with which the scheduler kills tasks in the job pool running over capacity.

Capacity scheduling [7]—In fair scheduling, the task slots are allocated to different jobs in a static manner. However, in real world scenario, it is possible that a job cannot fully utilize all the task slots while another concurrent job overloads its task slots. To solve this problem, capacity scheduling policy is proposed. It provides capacity guarantees for queues while providing elasticity for queues cluster utiliza-

[1]A MapReduce scheduler is a scheduler that schedules map and reduce tasks.

tion in the sense that unused capacity of a queue can be harnessed by overloaded queues that have a lot of temporal demand.

Delay scheduling [33]—In the previous scheduling policy, the main purpose is to ensure fair resource allocation between multiple jobs. The delay scheduling policy aims to maximize the overall performance of Hadoop platform. In MapReduce, the data associate with a task could be stored in either remote datanode or local datanode. It is much faster to read data from local datanode. In order to maximize the chance of local data access, in delay scheduling, when a node requests a particular task from a job, if the job is not able to assign local task, the scheduler skip that task and looking for next jobs. Obviously, delay scheduling policy may incur job starvation. In order to resolve this problem, proper precaution steps are compulsory required to avoid starvation effect. The delay scheduling improves problem of locality by asking jobs to wait for scheduling opportunity on a node with local data.

Besides the above four popular task scheduling policies, researchers have proposed many other policies, such as dynamic priority scheduling, deadline-based scheduling, and resource-aware scheduling. MapReduce is increasingly popular in large data set processing. There have been a lot of research works on its adaption and improvement [13, 26, 28, 32].

7.1.2 Adapting to Other Platforms

MapReduce scheduling has been extended to a great many of platforms, such as shared-memory multi-core platform, Cell broadband platform, GPU, FPGA and mobile platform. Phoenix [10] is a MapReduce framework on shared-memory multi-core architecture. Based on Phoenix, [18, 31] optimized the performance of MapReduce on multi-core platform. MapReduce frameworks [11, 23] are also proposed for Cell broadband engine architecture. Different from [11, 23] focused on MapReduce on asymmetric Cell-based clusters using a streaming approach while [11] only implement a MapReduce framework on a single Cell processor. Mars [11] harnesses the GPU computation power and high memory bandwidth to accelerate MapReduce frameworks, such as Hadoop. In this case, MapReduce applications are executed on both CPUs and GPUs. FPMR [27] is proposed for developers to create MapReduce programs on FPGA. Ref. [13] proposed a MapReduce framework on heterogenous mobile platform. For shared-memory multi-core, Phoenix [10] is implemented as a MapReduce framework includes a programming API and an efficient runtime system. However, Phoenix only performed well on small-scale systems with uniform access latencies. For large-scale NUMA systems, [31] optimized Phoenix using a multi-layered approach that comprises optimizations on the algorithm, implementation and OS interaction. MATE [18] is another MapReduce implementation extended from Phoenix which provides a high-level but distinct API.

7.1.3 Variations of MapReduce

Improving the performance of MapReduce has been a popular research issue. Assigning tasks to appropriate nodes is an efficient way to improve the performance of MapReduce. A lot of efficient MapReduce scheduling algorithms have been proposed to improve the performance of MapReduce in many scenarios. Fischer et al. [15] proposes an idealized mathematic model to evaluate the cost of task assignments and develops a flow-based algorithm to optimally assign tasks. Polo et al. [22] proposed an infrastructure aware MapReduce scheduler that monitors the tasks and evaluates the benefits of running each task on different nodes in real time. Based on the evaluation, the scheduler can decide the best distribution of tasks on nodes accordingly. Chen et al. [8] proposed a Tiled-MapReduce scheduling algorithm that partitions a large MapReduce tasks into a number of small sub-tasks and iteratively processes one sub-task at a time with efficient use of resources. Zaharia et al. [20] proposed a fair-sharing algorithm for a multi-user MapReduce system that arrange system resources (map/reduce task slots) for many users fairly. Zaharia et al. [33] introduced a delay scheduling algorithm. Aboulnaga et al. [1] proposed a MapReduce scheduling algorithm to minimize the execution time and improve the system resources utilization. The algorithm defines virtual machines (VM) and allocates the VMs to jobs, and to physical nodes. Sandholm et al. [24] designed a Dynamic Priority (DP) parallel task scheduler that allows users to control their allocated capacity by dynamically adjusting their budgets.

7.1.4 Existing Problem in Heterogeneous Environment

Because a MapReduce job is not completed until all the data is processed, the execution time of the job is decided by the last finished tasks (i.e., the weakest link effect). The scheduling policies introduce above work well in homogeneous environment, because every task can be processed in similar time. In heterogeneous environments, on the other hand, the execution time of a MapReduce job is seriously damaged by straggler tasks that run much slower than other tasks. This is mainly because workers on different nodes require various time in accomplishing even the same tasks due to their differences, such as capacities of computation and communication, architectures and memorizes.

One of the most popular solutions of this problem in MapReduce is launching backup tasks for straggler tasks on fast node. If a MapReduce scheduler launches a backup task γ_b for a straggler task γ, the small data set of γ is processed completely when either γ_b or γ finishes. In this case, if γ_b finishes before γ, the execution time of the job is reduced.

Although current MapReduce schedulers try to launch backup tasks for straggler tasks, they fail to detect straggler tasks correctly due to the wrong-estimated remaining time of all the tasks [16, 19]. The wrong detected straggler tasks cause at least two problems. First, launching backup tasks for these wrong straggler tasks cannot improve the performance of the MapReduce job since the real straggler tasks still prolong the execution time. Second, the backup tasks which are launched for the wrong straggler tasks waste system resources. The contention on the system resources even degrades the overall performance of the MapReduce job.

The wasting of system resources is one main problem of the backup strategy. Currently, MapReduce schedulers classify nodes into fast nodes and slow nodes, so that backup tasks can be launched on fast nodes. However, slow nodes can be further classified into *map slow nodes* and *reduce slow nodes* in a real system, since it is very possible that a node processes map tasks fast but processes reduce tasks slow and vice versa. We use map/reduce slow nodes to represent the nodes that execute map/reduce tasks slow than most of other nodes. The un-distinguishing between map slow nodes and reduce slow nodes wastes system resources. Let us take a reduce task γ that needs a backup task for example. Current MapReduce schedulers will not launch the backup task on a slow node N_s. However, if N_s is only a *map slow node*, launching the backup task of γ on N_s can utilize resources on N_s efficiently and improve the overall performance, since N_s can process reduce tasks fast.

7.2 Prior Solutions

There are two policies to detect straggler tasks: the least progress policy and the longest remaining time policy. For example, Hadoop [16] uses the least progress policy while LATE [19] uses the longest remaining time policy to detect straggler tasks. Both policies need to estimate the progress of every map/reduce task accurately. ParaTimer [21] is a time-oriented progress indicator for parallel queries that ensembles of MapReduce jobs. However, the indicator can only estimate the progress of SQL queries.

7.2.1 Least Progress Policy

As mentioned before, when a node has an empty task slot, Hadoop chooses a task to execute. It chooses tasks from one of three categories. First, any failed tasks are given highest priority. This is done to detect when a task fails repeatedly due to a bug and stop the job. Second, non-running tasks are considered. For maps, tasks with data local to the node are chosen first. Finally, Hadoop looks for a task to execute speculatively to speed up the straggler tasks.

In order to select speculative tasks, Hadoop monitors task progress using a *progress score* between 0 and 1. For a map task, the progress score is the fraction of input data read. For a reduce task, the execution is divided into three phases(*copy*, *sort*, and *reduce*), each of which accounts for 1/3 of the score. In copy phase, the task fetches map outputs; In sort phase, map outputs are sorted by key; In reduce phase, a user-defined function is applied to the list of map outputs with each key.

Hadoop looks at the average progress score of each category of tasks (maps and reduces) to define a *threshold* for speculative execution: when the progress score of a task is far less than the average for its category, it is marked as a straggler. Although a metric like progress rate would make more sense than absolute progress for identifying stragglers, the threshold in Hadoop works reasonably well in homogenous environments because tasks tend to start and finish in "waves" at roughly the same times and speculation only starts when the last wave is running.

However, the least progress policy performs poor in heterogeneous environment. Because the policy ranks candidates by locality, the wrong tasks may be chosen for speculation first. For example, if the average progress was 80% and there was a 2x slower task at 40% progress and a 10x slower task at 8% progress, then the 2x slower task might be speculated before the 10x slower task if its input data was available on an idle node.

7.2.2 Longest Approximate Time to End Policy

In order to solve the above problem, Longest Approximate Time to End (**LATE**) policy [19] is proposed. In LATE policy, the task that will *finish farthest into the future* is speculatively executed first, because this task provides the greatest opportunity for a speculative copy to overtake the original and reduce the job's response time.

In the policy, for each task, its progress rate (denoted by R) and its remain time to completion (denoted by TTE) are calculated in Eq. 7.1, where T is the amount of time the task has been running for and *PS* is the progress score of the task (collected int the same way as the least progress policy).

$$R = \frac{PS}{T}, TTE = \frac{1 - PS}{R} \tag{7.1}$$

The above calculation assumes that tasks make progress at a roughly constant rate. To really get the best chance of beating the original task with the speculative task, the policy only launch speculative tasks on fast nodes—not stragglers. To achieve this purpose, the scheduler does not launch speculative tasks on nodes that are below some threshold, *SlowNodeThreshold*, of total work performed (sum of progress scores for all succeeded and in-progress tasks on the node). This heuristic leads to better performance than assigning a speculative task to the first available node.

To handle the fact that speculative tasks cost resources, a threshold *Speculative-Cap* is given to limit the number of speculative tasks that can be running at once. Furthermore, a threshold *SlowTaskThreshold* that a task's progress rate is compared with to determine whether it is "slow enough" to be speculated upon. This prevents needless speculation when only fast tasks are running.

In summary, the LATE policy works as follows. If a node asks for a new task and there are fewer than *SpeculativeCap* speculative tasks running:

- Ignore the request if the node's total progress is below *SlowNodeThreshold*.
- Rank running tasks that are not currently being speculated by estimated time left.
- Launch a copy of the highest-ranked task with progress rate below *SlowTask-Threshold*.

7.2.3 Calculating Progress Score

Both the least progress policy and the LATE policy monitor the progress of every task using *progress score* (ranges from 0 to 1). In current MapReduce system, the execution of a map task comprises two phases and the execution of a reduce task comprises three phases as shown in Fig. 7.1. Therefore, the progress score of a task comprises from the progress score of every phase. Current MapReduce schedulers, such as Hadoop's scheduler and LATE, assume that $M1$, $M2$, $R1$, $R2$ and $R3$ are 1, 0, 1/3, 1/3 and 1/3 respectively.

By accumulating the progress score of different phases, the progress of a task can be calculated. The basis of calculating the progress score of a task is calculating the progress of a phase. The progress of a phase, denoted by PS_{phase}, can be calculated in Eq. 7.2. In the equation, M is the number of key/value pairs that have been processed in the phase and N is the overall number of key/value pairs that needed to be processed in the phase.

$$PS_{phase} = \frac{M}{N} \tag{7.2}$$

Let me introduce how to calculate the progress score of a task in more detail in the existing two policies. Take a task γ for example. If γ is a map task, since the first phase occupies the overall progress score, the progress score of γ is the progress $M1$. If γ is a reduce task and the first K phases of γ has finished, since each phase occupies 1/3 of the progress score, PS_γ is calculated by adding the progress score of the finished phases and the progress score of the current phase. Therefore the progress score of γ, denoted by PS_γ, is calculated in Eq. 7.3.

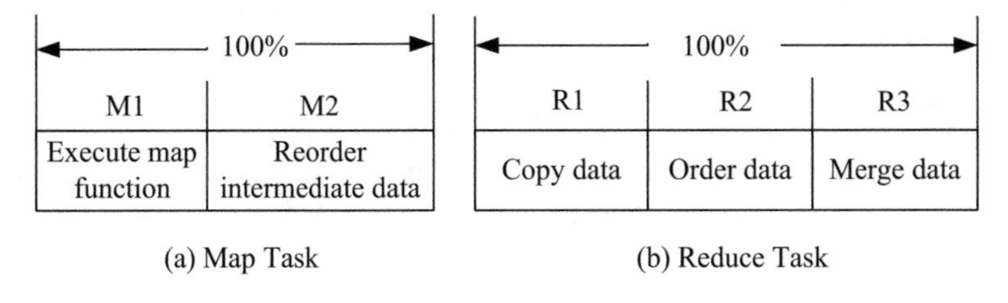

Fig. 7.1 Two phases of a map task and three phases of a reduce task

$$PS_\gamma = \begin{cases} PS_{phase} & \gamma \text{ is a map task,} \\ \frac{1}{3} \times K + \frac{1}{3} \times PS_{phase} & K \in (0, 1, 2), \gamma \text{ is a reduce task.} \end{cases} \quad (7.3)$$

For a MapReduce system with n running tasks ($\gamma_1, \gamma_2 ... \gamma_n$), the average progress score of the n running tasks, denoted by PS_{avg}, is calculated in Eq. 7.4.

$$PS_{avg} = \sum_{i=1}^{n} \frac{PS_i}{n} \quad (7.4)$$

Suppose task γ_j's progress score is PS_j and it has run T_j seconds ($j \in (1, 2, ..., n)$). If the least progress policy is used to detect straggler tasks, γ_j is a straggler task only when $PS_j \leq PS_{avg} - 20\%$.

On the other hand, if the LATE policy is used, the remaining time of all the n tasks needs to be calculated further. Then the scheduler chooses the tasks with the longest remaining time as straggler tasks. To calculate the remaining time of task γ_j, the progress rate of γ_j, denoted by PR_j, is calculated first in Eq. 7.5. Based on Eq. 7.5, the remaining time of γ_j, denoted by TTE_j, is calculated Eq. 7.6.

$$PR_j = \frac{PS_j}{T} \quad (7.5)$$

$$TTE_j = \frac{1.0 - PS_j}{PR_j} = T \times \frac{1.0 - PS_j}{PS_j} \quad (7.6)$$

7.2.4 Problems in Existing Solutions

In most cases, LATE policy works better than the least progress policy [19]. This is because the task with a small progress score does not always complete later than the task with a high progress score, especially in heterogeneous environment. For example, in a MapReduce system that has six tasks (γ_1, γ_2, ..., γ_6), suppose their progress scores are 0.7, 0.5, 0.9, 0.9, 0.9 and 0.9 respectively. We further suppose that they need 100, 30, 10, 10, 10 and 10 s to finish their work. In this case, $PS_{avg} = (0.7 + 0.5 + 0.9 * 4)/6 = 0.8$. The least progress policy classifies γ_2 to be a straggler task. However, γ_1 is the real straggler task since γ_1 needs more time to finish its work.

If MapReduce scheduler can accurately predict the real progress of each task, leveraging LATE policy, we can identify real straggler tasks. However, emerging MapReduce schedulers fail to calculate the progress score accurately. While $M1$, $M2$, $R1$, $R2$ and $R3$ vary across hardware settings and MapReduce applications in real system execution, they are constantly set to be 1, 0, 1/3, 1/3 and 1/3 respectively in emerging MapReduce schedulers.

Let me take a node with $R1 = 0.6$, $R2 = 0.2$ and $R3 = 0.2$ as an example to explain the poor progress prediction in emerging MapReduce schedulers. Suppose a reduce task γ has completed the first phase and has run T seconds on the node, the remaining time of γ is $T * \frac{1-0.6}{0.6} = 0.67T$ s. However, because $R1$, $R2$ and $R3$ are constantly seted to be 1/3 in emerging MapReduce schedulers, the calculated remaining time of γ is $T * \frac{1-1/3}{1/3} = 2T$ s. Based on the wrong remaining time of each task, LATE policy is not able to correctly identify real straggler tasks.

To this end, we propose a Self-Adaptive MapReduce (SAMR) scheduler that adjusts $M1$, $M2$, $R1$, $R2$ and $R3$ based on the historical values of them in the completed tasks. Based on the specific values of them for the current hardware features and application features, SAMR can estimate the progress scores of running tasks accurately, and hence can find real straggler tasks.

7.2.5 *Tarazu*

The above techniques try to identify straggler tasks and speed up their execution, so that improve the whole performance of an MapReduce application. Besides straggler tasks, there are two more key factors may result in the poor performance of an application: (1) MapReduce's built-in load balancing of Map computation results in excessive network communication and (2) the heterogeneity amplifies the load imbalance in Reduce computation. These factors extend beyond the issues of stragglers and speculative execution, which is the main part of this chapter.

In order to solve the above two pain points in MapReduce, Ahmad et al. [2] proposed Tarazu, an improved MapReduce scheduler, to optimize the performance of MapReduce on heterogeneous clusters. Tarazu consists of a *Communication-Aware Load Balancing of Map computation* (CALB) policy that regulates the use of remote Map tasks based on whether Map or Shuffle is likely to be in the critical path, *a Communication-Aware Scheduling of Map computation* (CAS) policy that spread out remote map task traffic over time, and a *Predictive Load Balancing of Reduce computation* (PLB) policy that balances reduce tasks across heterogeneous nodes. Figure 7.2 shows the general overview of the three policies.

7.2.5.1 Communication-Aware Load Balancing of Map (CALB)

CALB is designed based on the key observation that due to the overlap between Map computation and Shuffle, either the Shuffle or the Map computation is in the critical path, depending upon the MapReductions Shuffle load and the cluster hardware characteristics.

If the Shuffle is critical, CALB switches to no-steal mode where CALB prevents task stealing for most of the Map phase, preventing further aggravation of the Shuffle traffic and increasing the chances of tasks being executed locally. When the Shuffle traffic falls below a threshold, CALB allows task stealing to load-balance any remaining tasks (naturally, faster nodes steal work from slower nodes). Intuitively, in this case, for most of the Map phase, fast nodes do not steal tasks, allowing the slow

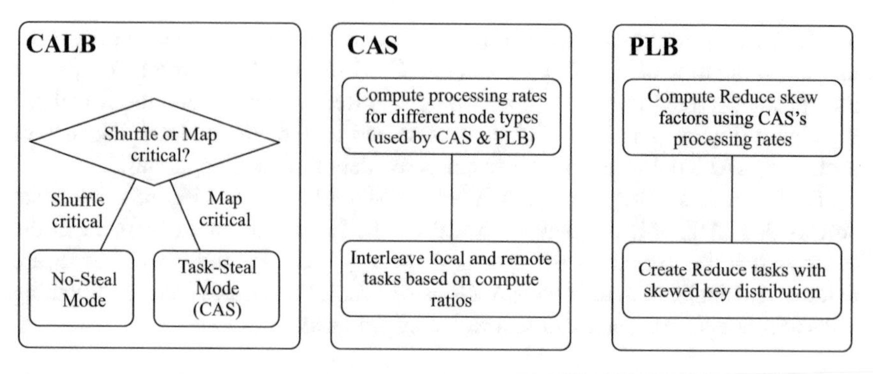

Fig. 7.2 The general overview of the CALB, CAS, and PLB policies in Tarazu [2]

nodes to run the Map tasks locally. At the end of the Map phase when the Shuffle ends, fast nodes steal a few tasks from remote slow nodes.

On the other hand, if the Map computation is critical, CALB continues in the task-steal mode. The number of remote tasks needed and when to schedule them will be handled in CAS, which is discussed later. In the Map phase, fast nodes steal tasks from slow nodes. CALB hides the resultant fast nodes remote task traffic under the slow nodes computation. By load-balancing Map computation, CALB shortens the critical path in this case.

In order to identify whether the Shuffle or the Map computation is critical, Ahmad et al. [2] believed that shuffle is critical when Map tasks complete their computation at a faster rate than their communication to Reduce tasks. Therefore, CALB Leverages the number of Map tasks that have completed their computation ($Computed_i$) and the number that have completed their communication ($Communicated_i$) for each node i to identify the critical phase. In this case, suppose there are overall N nodes, $|\sum_{i=1}^{N} Computed_i - \sum_{i=1}^{N} Communicated_i|$ indicates the extent to which the Shuffle lags the Map computation. If it increases over time, it implies that the Shuffle is likely to be critical.

7.2.5.2 Communication-Aware Scheduling of Map (CAS)

While CALB decides whether to allow remote task stealing would be beneficial for an MapReduce application, CAS determines how many remote tasks are needed and when to execute them in the task-steal mode.

In emerging MapReduce implementation, tasks are stolen across nodes at the end of the Map phase, creating a surge of traffic that may results in poor performance. In order to avoid this problem, CAS spreads out the task stealing across nodes (during initial part of the Map phase, and in CALBs task-steal mode) by interleaving them with local task execution. In addition to avoiding the bursty traffic, CAS has other benefits: (1) By interleaving remote tasks with local tasks in the Map phase, CAS achieves better overlap between remote task communication and local task computation on both fast nodes and slow nodes. (2) The remote tasks read input data faster by avoiding bursts. These benefits shorten Map computation, which is the critical path relevant for CAS.

As described before, CAS is designed to steal remote tasks throughout the Map phase when the overall number of remote tasks is not known. In order to fulfill this requirement, Tarazu measures the average execution time of map task for each node type in the heterogeneous cluster and compute the ratios of the execution time for each pair of node types as in Eq. 7.7. In the equation, T_i is the time of a map task on a faster node of type i and T_j is the time of a map task on a slower node of type j.

$$mapRatio_{i,j} = \frac{T_j}{T_i} \tag{7.7}$$

CAS uses these ratios to determine the number of remote tasks to be moved from one node to another. The larger the ratio $mapRatio_{i,j}$ is, the more remote tasks are allowed to be stolen from node of type j to node of type i. Because current MapReduce implementations already track the identity of the pair of source and destination nodes involved in task stealing, CAS can apply the specific pair's $mapRatio$.

7.2.5.3 Predictive Load Balancing of Reduce (PLB)

While CALB and CAS optimize the Map phase, PLB achieves better load balance in the Reduce phase by skewing the intermediate key distribution among the Reduce tasks based on the type of the node on which a Reduce task runs. While current implementations create as many bins per Map task as there are Reduce tasks, PLB creates more hash bins (by a factor of *binMultiplier*) as the number of Reduce tasks to achieve the skew (e.g., binMultiplier = 4). PLB uniformly distributes keys to the bins and then assigns as many bins to each Reduce task on node i as is dictated by the skew factor, $reduceSkewFactor_i$. For instance, if a fast node is three times as fast as a slow node, then $reduceSkewFactor_i$ for the fast node is 3 and that for the slow node is 1. Therefore, a Reduce task on a fast node gets three times as many bins as a Reduce task on a slow node.

To implement assigning multiple bins per Reduce task, Ahmad et al. [2] modified the MapReduce implementation to allow multiple sends from a Map task to a Reduce task (the baseline implementation assigns and sends only one bin from a Map task to a Reduce task). Note that although more hash bins per Map task are created in Tarazu than the baseline, applications have the same number of Reduce tasks in Tarazu as the baseline.

By integrating CALB, CAS and PLB together in Tarazu, the workload can be balanced across the nodes in heterogeneous clusters while reducing the burst network traffic. While Tarazu improves the performance of MapReduce on heterogeneous clusters by balancing the workloads, we further introduce the other categories of techniques: accelerating straggler tasks. Tarazu is orthogonal to the techniques introduced in the following of this chapter. They are able to be integrated to achieve better performance.

7.3 Self-adaptive MapReduce Scheduling

This section presents SAMR, a Self-Adaptive MapReduce scheduler. In this section, we first overview the design of SAMR. Then, we present the historical-based strategy that tunes runtime parameters used by SAMR automatically. After that, we present the detailed algorithms for detecting straggler tasks, detecting slow nodes, and selecting appropriate backup node for straggler tasks.

7.3.1 Overview of SAMR

Fig. 7.3 presents the design overview of our self-adaptive MapReduce scheduler, SAMR. As shown in the figure, SAMR monitors the progress of all the active tasks. Based on the progress of each task, SAMR identifies straggler tasks and boosts the execution of straggler tasks through launching a backup task for each straggler task on a fast node.

In more detail, SAMR executes a MapReduce job in the following ways. First, when SAMR receives a MapReduce job W, SAMR identifies the category of W. If W is an instance of application P, every worker reads in the $M1$, $M2$, $R1$, $R2$ and $R3$ when executing application P from its local node. During the execution of W, the values of $M1$, $M2$, $R1$, $R2$ and $R3$ on each node are tuned dynamically according to their actual values. Second, based on the dynamic-tuned $M1$, $M2$, $R1$, $R2$ and $R3$, SAMR can compute progress scores of tasks more accurate, which is the basis of straggler task detecting. Meanwhile, SAMR detects slow nodes according to the

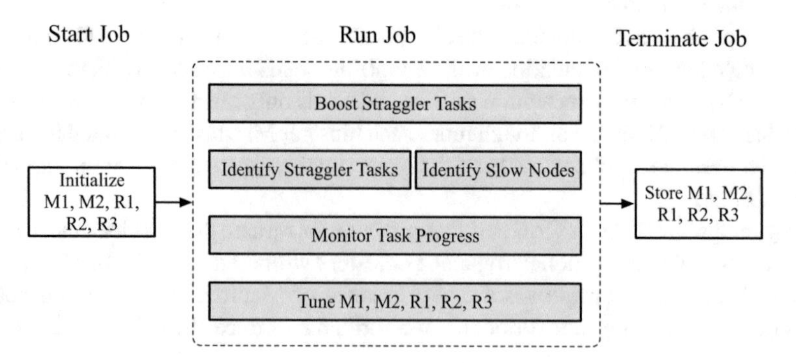

Fig. 7.3 Design of SAMR

average progress rates of map tasks and reduce tasks on every node (to be described shortly). When a straggler task t is detected, SAMR launches a backup task for it. After all the data sets have been processed, SAMR terminates the MapReduce job and reports the final result. Lastly, every node stores the updated $M1$, $M2$, $R1$, $R2$ and $R3$ for application P, so that future instance of P can benefit from the accurate historical information.

7.3.2 *Tuning Phase Weights*

In heterogeneous Cloud platform, workers on different nodes operate at various speeds. It is not accurate to use the same phase weight configuration (i.e., $M1$, $M2$, $R1$, $R2$ and $R3$) for all the workers. To this end, SAMR tunes the weights of every phase in map task and reduce task on each individual node using a history-based auto-tuning strategy. In this strategy, a worker p reads in the historical weights of $M1$, $M2$, $R1$, $R2$ and $R3$ from the corresponding node to be its default weight configuration when the worker was started. Once the worker p completes a map task, its $M1$ and $M2$ are updated. Similarly, once p completes a reduce task, $R1$, $R2$ and $R3$ are updated. In this way, each worker has its own phase weight configuration that reflects the actual task processing character on the corresponding node.

In more detail, SAMR tunes the weight of each phase ($M1$, $M2$, $R1$, $R1$ and $R3$) as follows. Suppose the current weight of a phase is V_{old}, and the actual weight of the phase in a newly-completed task is V_{cmpl}. The new weight of the phase, denoted by V_{new}, is calculated in Eq. 7.8, in which HP shows how badly the history information impacts V_{new}.

$$V_{new} = V_{old} \times HP + V_{cmpl} \times (1 - HP) \tag{7.8}$$

Observed from Eq. 7.8, if HP is too large (close to 1), V_{new} highly depends on V_{old}. In this case, V_{new} is not able to capture the up-to-date features of the current running tasks. On the other hand, if HP is too small (close to 0), the appropriate value of the weight may be destroyed by random factors, since V_{cmpl} is likely to be influenced by random events.

It is worth noting that there is not any additional communication when a worker reads and updates historical information, since every worker reads and writes historical information from local node. Therefore, SAMR is scalable.

7.3.3 *Calculating Progress Score*

Based on the timely-undated weight of each phase, SAMR is able to calculate the progress score of each task precisely. When SAMR executes a MapReduce job, it computes the progress score of every active task periodically (e.g., every 100 ms). Take an active task γ for example. Suppose the worker has processed K phases of

γ. Equations 7.9 and 7.10 compute the progress score of γ when it is a map task or a reduce task, respectively. In the two equations, PS_{phase} is the progress score of the current phase and can be computed in Eq. 7.2.

$$\text{For map task: } PS_\gamma = \begin{cases} M1 \times PS_{phase} & \text{if } K = 0, \\ M1 + M2 \times PS_{phase} & \text{if } K = 1. \end{cases} \tag{7.9}$$

$$\text{For reduce task: } PS_\gamma = \begin{cases} R1 \times PS_{phase} & \text{if } K = 0, \\ R1 + R2 \times PS_{phase} & \text{if } K = 1, \\ R1 + R2 + R3 \times PS_{phase} & \text{if } K = 2. \end{cases} \tag{7.10}$$

7.3.4 Identifying Straggler Task

Based on the accurate progress score of every active task, SAMR is able to identify the real straggler tasks that may seriously damage the job's performance. In our design, a task γ is considered to be a straggler task only when both of the following two constraints are satisfied.

- Its data processing speed is much slower than other tasks' data processing speed.
- It is one of the tasks with the longest remaining time.

If the first constraint is not satisfied, even if the task γ has the longest remaining time, it should be be treated as a straggler task. For instance, if γ is a newly launched task on a fast node, it is quite possible that γ has the longest remaining time although the data processing rate of γ is fast. In this case, there is no benefit to launch a backup task for γ because the backup task is not able to complete before γ, and it should not be treated as a straggler task. Otherwise, if only the first constraint is satisfied but the second constraint is not satisfied, task γ is not the task that would complete lastly. In this case, it is not necessary to launch a backup task for γ.

In order to identify whether task γ is a slow task, we compare the data processing rate of γ with the average data processing rate of the same type of tasks in the whole job. Let PS_γ and PS_{avg} represent the data progressing rate of γ and the average data progressing rate of the job. Only when γ's data progressing rate PS_γ fulfills Eq. 7.11, it is considered to be a slow task. In the equation, $Task\_Cap$ is the threshold that identifies slow tasks.

$$PR_\gamma < (1.0 - Task\_Cap) \times PR_{avg} \tag{7.11}$$

Observed from Eq. 7.11, if Task_Cap is too small (close to 0), SAMR will classify some fast tasks into slow tasks. On the other hand, if Task_Cap is too large (close to 1), SAMR will classify some slow tasks into fast tasks. In the evaluation section, we show the performance of SAMR with different Task_Cap.

After all the current slow tasks are identified, SAMR computes the remaining time to complete of each of the slow tasks using Eq. 7.6. SAMR chooses the slow tasks with the longest remaining time to be the straggler tasks adopting the LATE policy.

Furthermore, in order to handle the fact that backup tasks of straggler tasks cost resources, SAMR limits the number of straggler tasks. Therefore, a cap on the number of straggler tasks (i.e., the number of backup tasks since SAMR only allows one backup task for each straggler task), denoted by $Strag\_Cap$, is used. Suppose the number of the overall running tasks is $Task\_Num$, the up-bound of the number of straggler tasks, denoted by $Strag\_UB$, is $Strag\_Cap \times Task\_Num$. If Strag_Cap is too small (close to 0), some real straggler tasks is overlooked by SAMR. On the contrary, if Strag_Cap is too large (close to 1), too many tasks could be identified to be straggler tasks. The backup tasks for these straggler tasks cost a lot of system resources and may degrade the overall performance in consequence. Algorithm 11 gives the algorithm of detecting straggler tasks in SAMR.

Algorithm 11 Straggler tasks detecting algorithm

```
1: DetectStragglerTask() {
2:   While (the job is running) {
3:     Every worker computes the progress rate of every active tasks on it ;
4:     SAMR computes the average progress rate of all the running tasks ;
5:     Every worker identifies slow tasks according to Eq. 7.11 ;
6:     Every worker reports the list of its slow active tasks ;
7:     Computes the remaining time for all the slow tasks according to Eq. 7.6;
8:     Sorts the slow tasks in the descending order of their remaining time ;
9:     Calculate the cap of the number of straggler tasks, Strag_UB ;
10:     If (the number of slow tasks ≤ Strag_UB)
11:       All the slow tasks are considered to be straggler tasks ;
12:     else
13:       Strag_UB slow tasks with the longest remaining time as straggler tasks ;
14:     Inserts all the straggler tasks into straggler map/reduce task list ;
15:     usleep(100000) ; //SAMR detects straggler tasks every 100 ms
16:   }
17: }
```

7.3.5 Identifying Slow Node

In heterogenous environments, different nodes have different CPU, memories and I/O devices. The difference leads to different rate in executing map tasks and reduce tasks. During the execution of a MapReduce job, if SAMR launches a backup task for a straggler task on a slow node, there is not any performance improvement since the backup task will be finished even later than the original straggler task on a slow node. To improve performance and decrease response time of a job, SAMR does not launch backup tasks on slow nodes. However, it is also possible that some nodes execute map tasks slow but execute reduce tasks fast. To increase the resource utilization while ensuring the performance, SAMR classifies slow nodes into map slow nodes and reduce slow nodes further. The backup tasks of straggler map tasks can also be launched on reduce slow nodes besides fast nodes and vice versa.

To detect slow nodes in the system, SAMR uses the average progress rate of the running map/reduce tasks on a node to represent the map/reduce task progress rates of the node. The nodes with the smallest map/reduce task progress rate are map/reduce slow nodes. Given a node Φ with M map tasks and R reduce tasks. The map/reduce task progress rates of Φ, denoted by MR_{Φ} and RR_{Φ}, are calculated in Eq. 7.12, where PR_i is the progress rate of the ith map/reduce task.

$$\begin{cases} MR_{\Phi} = \sum_{i=1}^{M} PR_i/M, \\ RR_{\Phi} = \sum_{i=1}^{R} PR_i/R. \end{cases} \tag{7.12}$$

For node Φ, if $MR_{\Phi} < (1 - Node\_Cap) \times MR_{avg}$, it is a map slow node. If $RR_{\Phi} < (1 - Node\_Cap) \times RR_{avg}$, it is a reduce slow node. MR_{avg} and RR_{avg} are the average map/reduce tasks progress rate of all the nodes. $Node\_Cap$ is the threshold that identifies slow nodes.

Therefore, if *Node_Cap* is too small (close to 0), SAMR will classify some fast nodes into slow nodes. On the other hand, if *Node_Cap* is too large (close to 1), SAMR will classify some slow nodes into fast nodes.

To limit the number of slow nodes, a cap on the number of slow nodes, denoted by SN_Cap, is introduced in SAMR. Suppose the number of the overall node is $Node\_Num$. The up-bound of the number of slow map/reduce nodes is $SN\_Cap \times Node\_Num$.

7.3.6 Boosting Straggler Task

Launching backup tasks for straggler tasks is one of the most popular method to boost the straggler tasks and improve the performance of a MapReduce job. Since SAMR is able to identify straggler tasks and map/reduce slow nodes accurately in a timely manner, SAMR can simply launch backup tasks for straggler tasks. When a node Φ is free, it first tries to obtain a new task that never been executed before. If there is not any new task, Φ checks whether it is a map or reduce slow node. If Φ is not a map slow node, Φ launches a backup task for a straggler map task. If Φ is not a reduce slow node, Φ launches a backup task for a straggler reduce task. Algorithm 12 shows the detailed algorithm used to obtain a new task when node Φ is free.

In addition, SAMR also limits the number of backup tasks since backup tasks cost system resources. As mentioned in Sect. 7.3.4, the up-bound of the number of backup tasks is $Strag\_Cap \times Task\_Num$.

Algorithm 12 Algorithm used to obtain a new task by node Φ

```
1: While (the job is still running) {
2:   Φ tries to get a new task that never been executed by other nodes before ;
3:   If (succeed)
4:     Φ starts to execute the obtained task ;
5:   Else {
6:     Φ checks whether it is a map/reduce slow node ;
7:     If (Φ is not a map slow node) {
8:       Φ tries to pop a straggler map task from the straggler map task list ;
9:       Φ launches a backup task for the straggler map task ;
10:    }
11:    If (Φ is not a reduce slow node) {
12:      Φ tries to pop a straggler reduce task from the straggler reduce task list ;
13:      Φ launches a backup task for the straggler reduce task ;
14:    }
15:  }
16: }
```

7.4 Implementation of SAMR

Figure 7.4 shows the general architecture of the proposed SAMR scheduler. As shown in the figure, SAMR uses *straggler map task pool* and *straggler reduce task pool* to record straggler map tasks and straggler reduce tasks respectively. When a node tries to launch a backup task for either a straggler map task or a straggler reduce task, the straggler task with the longest remaining time is popped out from the corresponding task pool. In this way, SAMR always launches backup task for the straggler task which prolongs the execution time most serious first.

In SAMR, every node records the weights of every phase in a map task and a reduce task (i.e., $M1$, $M2$, $R1$, $R2$ and $R3$) for every history application, which partly reflect the execution features of tasks on the node. The weights can be stored in various formats, e.g., XML or Jason Format. For easy maintaining, in our current implementation, SAMR stores the weights in XML format, as shown in Fig. 7.5. When a job is submitted, SAMR first identifies it is an instance of which application. After that, every node uses an XML parser to search for the stored $M1$, $M2$, $R1$, $R2$ and $R3$ for the application. If they are found, the XML parser read in the stored weights and takes them as the default values of $M1$, $M2$, $R1$, $R2$ and $R3$ in the current execution. Otherwise, $M1$, $M2$, $R1$, $R2$ and $R3$ are configured to be 1, 0, 1/3, 1/3, 1/3 by default respectively. They will be tuned to according to the statistics collected at runtime.

In SAMR, all the nodes prefer executing unprocessed tasks from the *unprocessed task pool* rather than launching backup tasks. Due to the large data set of tasks, all the nodes prefer to execute tasks whose data set is stored on local node.

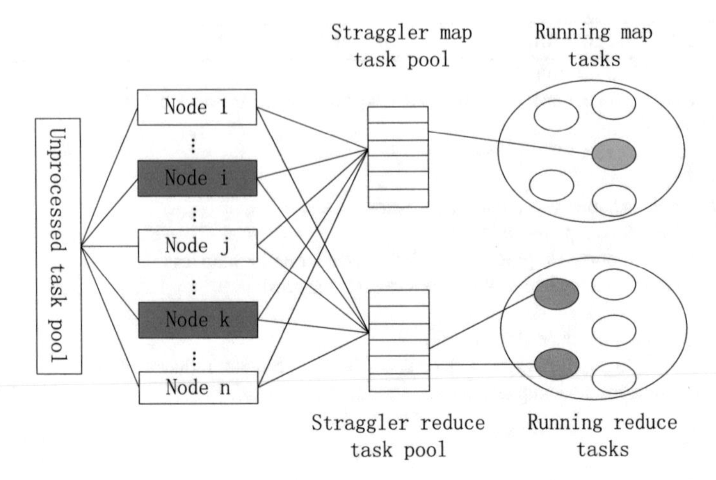

Fig. 7.4 Architecture of SAMR. Nodes can obtain tasks from both the unprocessed task pool and the straggler task pool. The *unprocessed task pool* stores all the unprocessed tasks. *Straggler map task pool* and *straggler reduce task pool* store the straggler map and reduce tasks. The filled squares are map/reduce slow nodes, and the filled cycles are either slow map tasks or slow reduce tasks respectively

```
<SAMR>
    <App>
        <NAME> APPLICATION-1 </NAME>
        <MAP> <M1>0.8</M1><M2>0.2</M2></MAP>
        <REDUCE> <R1>0.59</R1><R2>0.19</R2><R3>0.22</R3></REDUCE>
    </App>
    <App>
        <NAME> APPLICATION-2 </NAME>
        <MAP> <M1>0.45</M1><M2>0.55</M2></MAP>
        <REDUCE> <R1>0.55</R1><R2>0.15</R2><R3>0.3</R3></REDUCE>
    </App>
    ...
</SAMR>
```

Fig. 7.5 An example of $M1$, $M2$, $R1$, $R2$ and $R3$ that are recorded in XML format

7.5 Performance Evaluation

In this section, we compare the performance of emerging techniques used to boost straggler tasks in MapReduce job.

7.5.1 Experimental Setup

In the experiment, we use a small cluster that consists of five identical personal desktops. To build heterogenous environment, we have installed different number of

Table 7.1 Hardware configuration

	VMs per physic machine	Num of physic machines	Data write rate (MB/s)
Fast setting	1	2	2.87
	2	3	1.4
	Bare linux	1	3.43
Slow setting	1	1	2.87
	2	2	1.4
	2	1 slow machine	1.34
	Bare linux	1	3.43

virtual machines on the five homogenous physical machines. Each virtual machine has 1GB RAM and runs Linux 2.6.24. We compare SAMR with the default Hadoop scheduler and LATE [19], the state-of-the-art technique used to boost straggler tasks. For fairness of comparison, we implement both SAMR and LATE schedulers on Hadoop 0.19.1.

In order to evaluate the performance of Hadoop, LATE and SAMR in various hardware scenarios, as shown in Table 7.1, we simulate two heterogenous environments. Note that, we run a CPU-intensive program on one of the computers that have two virtual machines to simulate two extremely slow nodes in the slow setting. We choose two classic benchmarks, *Sort* and *WordCount*, to evaluate the performance of Hadoop, LATE and SAMR. For each test, every benchmark is run ten times and the average execution time is used as the result.

7.5.2 Performance

Because the slow setting in Table 7.1 provides more heterogeneity, in this section we report the performance of Hadoop, LATE and SAMR by evaluating them on the experimental platform with the slow setting. Experiments on the fast setting show similar results.

Figure 7.6 shows the performance of *Sort* and *WordCount* in SAMR, Hadoop and LATE scheduler. Observed from the figure, compared with Hadoop, SAMR significantly improves the performance of *Sort* and *WordCount*, with the performance gain up to 37% for *Sort* and up to 16% for *WordCount*. On the other hand, LATE can also slightly improve the performance of *Sort* and *WordCount*, with the performance gain up to 9% for *Sort* and up to 10.1% for *WordCount*. In this experiment, the performance results of SAMR are collected with the best configured parameters (i.e., HP, Task_Cap, Node_Cap, SN_Cap and Strag_Cap). We will describe the way to choose their appropriate values shortly.

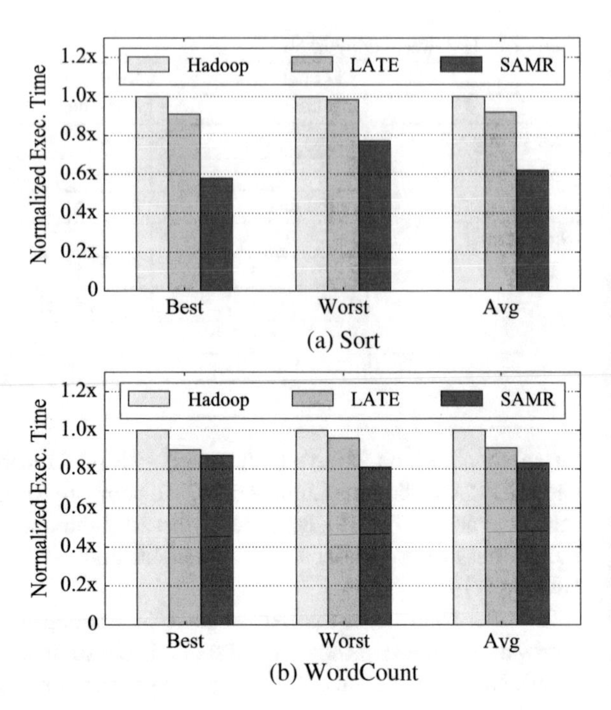

Fig. 7.6 Performance of *Sort* and *WordCount* with the slow setting when they are scheduled with Hadoop, LATE and SAMR

As shown in Fig. 4.12, both *Sort* and *WordCount* achieve a slightly better performance in LATE compared with Hadoop. The performance gains origin from the longest approximate time to end strategy in launching backup tasks [19]. Because SAMR is able to estimate the progress of tasks accurately, it can detect straggler tasks more accurate. Therefore, *Sort* and *WordCount* achieve better performance in SAMR compared with LATE.

7.5.3 *Effectiveness of Speculative Execution and Weight Tuning*

As mentioned before, emerging MapReduce schedulers employ speculative execution, with which the scheduler launches backup tasks for straggler tasks. In addition, SAMR tunes the weights of $M1$, $M2$, $R1$, $R2$ and $R3$ dynamically.

In order to evaluate the effectiveness of speculative execution and the weight tuning technique in SAMR, we compare the performance of the benchmarks when they are scheduled by the default *Hadoop*, *Hadoop-ns* (Hadoop without speculative execution) and *Hadoop-wt* (Hadoop with weight turning) on both the fast setting and the slow setting. Figure 7.7 shows the performance of *Sort* scheduled by the three schedulers on the two settings respectively. Experiment on *WordCount* shows similar results. From the figure we can see that the speculative execution is able to

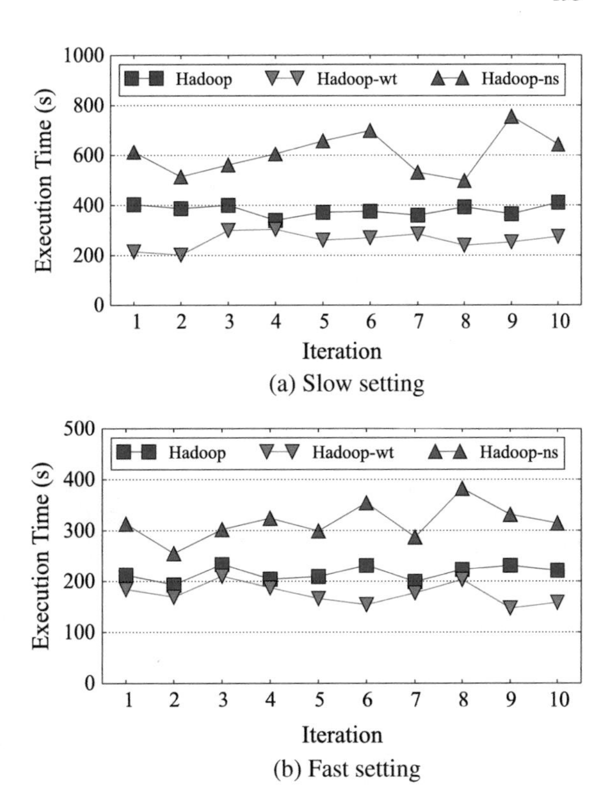

Fig. 7.7 Execution time of *Sort* on the slow setting and the fast setting when it is scheduled with Hadoop, Hadoop-ns, and Hadoop-wt

significantly improve the performance of MapReduce applications while enhancing the stability of execution time.

To demonstrate the effectiveness of the weight tuning technique proposed in SAMR, Table 7.2 lists the recorded values and the real values of $M1$, $M2$, $R1$, $R2$ and $R3$ on every node. Observed from the table, for map tasks, the difference between real values and the recorded values are less than 5%. On the other hand, for reduce tasks, in most cases, the difference between real values and the recorded value are less than 10%. Both the recorded values and the real values are far from the constant values of them employed in Hadoop's scheduler and LATE scheduler (i.e., 1, 0, $\frac{1}{3}$, $\frac{1}{3}$ and $\frac{1}{3}$). Based on the accurate weights of different phases, SAMR is able to detect the actual straggler tasks. Therefore, SAMR can improve the performance of MapReduce applications in heterogeneous environments.

7.5.4 Parameter Selection in SAMR

SAMR uses five parameters (i.e., HP, Task_Cap, Node_Cap, SN_Cap and Strag_Cap) to configure the scheduler for different hardware architecture and different

Table 7.2 The recorded/real values of $M1$, $M2$, $R1$, $R2$ and $R3$.

	Map task		Reduce task		
	M1	M2	R1	R2	R3
Node1	0.8/0.78	0.2/0.22	0.59/0.62	0.19/0.23	0.22/0.15
Node2	0.77/0.77	0.23/0.23	0.46/0.42	0.06/0.03	0.48/0.55
Node3	0.75/0.66	0.25/0.34	0.44/0.40	0.43/0.45	0.24/0.15
Node4	0.74/0.77	0.26/0.23	0.62/0.64	0.13/0.06	0.25/0.32
Node5	0.81/0.82	0.19/0.18	0.43/0.44	0.14/0.04	0.43/0.52
Node6	0.73/0.77	0.27/0.23	0.51/0.53	0.19/0.12	0.30/0.35
Node7	0.71/0.67	0.29/0.33	0.51/0.50	0.11/0.06	0.38/0.44
Node8	0.79/0.78	0.21/0.22	0.46/0.41	0.13/0.48	0.41/0.11

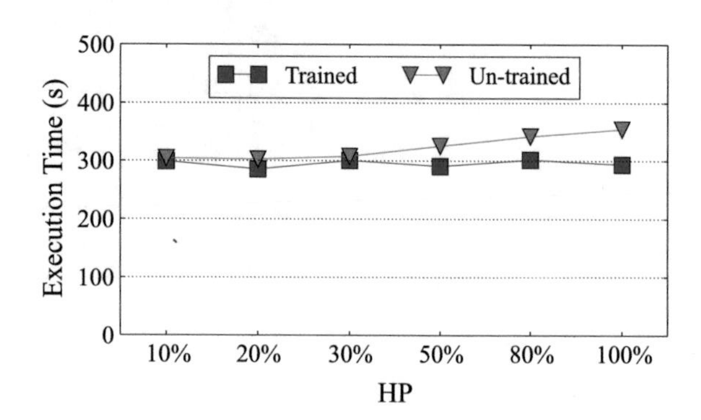

Fig. 7.8 Performance of *Sort* with different HP in the trained and untrained scenarios

applications. In order to find appropriate values for the parameters, we tune one parameter while fixing all the other parameters. According to the runtime algorithm of SAMR that is proposed in Sect. 7.3.1, we evaluate the performance of *Sort* when it is scheduled with SAMR having different HP, Task_Cap, Node_Cap, SN_Cap and Strag_Cap. Experiment on *WordCount* shows similar results.

Figure 7.8 shows the performance of *Sort* with different HP in two scenarios: untrained scenario and trained scenario. HP is the percentile of the recorded $M1$, $M2$, $R1$, $R2$ and $R3$ in their new values as defined in Sect. 7.3.2. We construct the trained scenario by executing *Sort* for two times before the current execution and construct the untrained scenario by setting the recorded $M1$, $M2$, $R1$, $R2$ and $R3$ to $1, 0, \frac{1}{3}, \frac{1}{3}$ and $\frac{1}{3}$ manually just like the assumption in Hadoop and LATE scheduler.

From the figure we can see that the value of HP does not affect the performance of *Sort* too much in the trained scenario. However, the performance of *Sort* degrades with the increasing of HP in the untrained scenario. The high and static performance of *Sort* in the trained scenario is resulted from the well-trained value of $M1$, $M2$, $R1$, $R2$ and $R3$. Figure 7.8 also suggests to use small HP if an application is executed for the first time. In this way, the value of $M1$, $M2$, $R1$, $R2$ and $R3$ can be tuned based on the current execution rapidly. Note that, if HP equals to 100%, the values of $M1$, $M2$, $R1$, $R2$ and $R3$ in the current job equal to the recorded value. In this

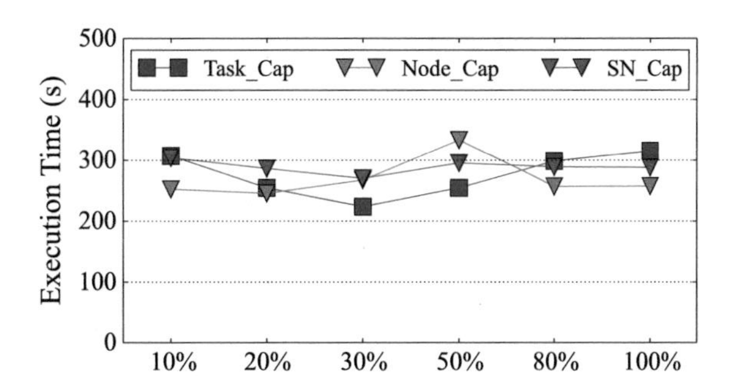

Fig. 7.9 Performance of *Sort* with different Task_Cap, Node_Cap and SN_Cap

case, SAMR scheduler is same to the LATE scheduler in the untrained scenario. In the following experiments, SAMR fixes HP to be 20%.

Figure 7.9 shows the performance of *Sort* with different Task_Cap, Node_Cap and SN_Cap in the slow setting. From the figure we can see that the best values of *Task_Cap*, *Node_Cap* and *SN_Cap* are 30%, 20%, 30% respectively.

Task_Cap is the percentile of speed below which a task will be considered too slow to be a slow task as defined in Sect. 7.3.4. As shown in Fig. 7.9, *Sort* gains best performance when Task_Cap is 30%. Deduced from Eq. 7.11, the smaller Task_Cap is the more tasks are classified to slow tasks. Therefore, if Task_Cap is smaller than 30%, some fast tasks are classified to slow tasks and even straggler tasks. In this case, the launching of backup tasks for these wrong-classified tasks consume a lot of system resources, so the overall execute time is prolonged. On the other hand, if Task_Cap is larger than 30%, some slow tasks and even straggler tasks are classified to fast tasks and none backup tasks are launched for them. These slow tasks will prolong the execute time as well.

Node_Cap is the percentile of speed below which a node will be considered too slow to be a map/reduce slow node as defined in Sect. 7.3.5. As shown in Fig. 7.9, *Sort* gains best performance when Node_Cap is 20%. If Node_Cap is small than 20%, some fast nodes are treated as map/reduce slow nodes by fault. In this case, the computing power of these wrong-classified nodes cannot be used to improve the performance by executing backup tasks for straggler tasks. On the other hand, if Node_Cap is larger than 20%, some map/reduce slow nodes are classified to fast nodes by fault. In the case, backup tasks may be launched on these slow nodes. Since the backup tasks on slow nodes will be finished later than the original straggler tasks, the overall execute time cannot be shorten.

SAMR uses SN_Cap to configure the maximum number of slow nodes in a cluster. As shown in Fig. 7.9, *Sort* gains best performance when SN_Cap is 30%. SN_Cap is useful if Node_Cap is not configured appropriately, because SN_Cap limits the number of slow nodes. SN_Cap guarantees that there are not too many nodes are classified to slow nodes. If SN_Cap is smaller than 30%, some map/reduce slow nodes may be classified to fast nodes when Node_Cap is too small (e.g., smaller than 20%). On the other hand, if SN_Cap is larger than 30%, some fast nodes may be

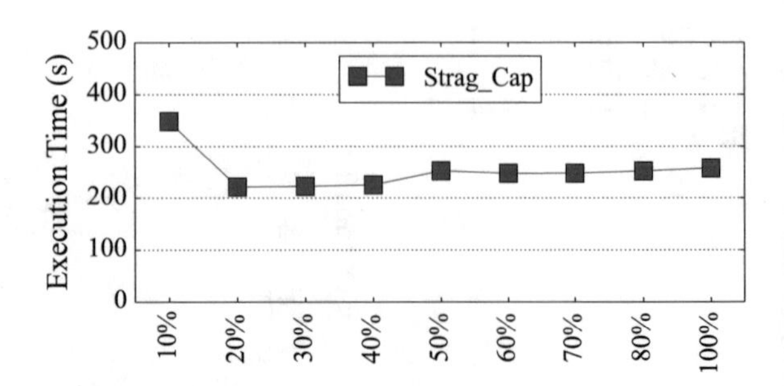

Fig. 7.10 The performance of *Sort* with different Strag_Cap

classified to slow nodes when Node_Cap is too large (e.g., larger than 30%). The wrong classification may result in the poor performance of SAMR.

SAMR uses Strag_Cap to configure the maximum number of backup tasks. As shown in Fig. 7.10, *Sort* gains the best performance when Strag_Cap equals to 30%. If Strag_Cap is smaller than 20%, SAMR is not able to launch backup tasks for all the straggler tasks due to the small number of backup tasks. On the other hand, if Strag_Cap is larger than 0.2, too many backup tasks will consume a large amount of system resources, so the execution time of *Sort* is prolonged as well.

After a series of experiments, the best parameters for SAMR are: HP = 20%, Task_Cap = 30%, Node_Cap = 20%, SN_Cap = 30%, Strag_Cap = 20% in our test bed for *Sort*. These parameters must be re-specified for any new cluster.

7.6 Summary

Traditional MapReduce schedulers for Cloud computing suffer from poor performance in heterogeneous environment, because they are not able to identify the real straggler tasks due to the poorly estimated progress scores of the active tasks. In order to address this problem, we have designed and implemented SAMR: a Self-Adaptive MapReduce scheduler that tunes the weight of each phase of a map task and a reduce task automatically based on historical statistics. Furthermore, SAMR classifies slow nodes into map slow nodes and reduce slow nodes. In this way, SAMR can launch backup tasks for reduce straggler tasks on map slow nodes and vice versa. Experimental results demonstrate that SAMR can achieve up to 37% performance gain compared with Hadoop, and up to 16% performance gain compared with LATE scheduler.

7.6.1 Chapter Highlights

The following highlights of this chapter could be of your interest:

- We have proposed a history-based technique that automatically updates the weights of different phases in a map/reduce task. Based on the accurate weights, the progress of each active task can be precisely estimated.

- Based on the precise estimated progress of each active task, we have proposed SAMR that is able to identify the actual straggler tasks that would significantly degrade the performance of a MapReduce job.
- We have proposed a technique that classifies slow nodes into map slow nodes and reduce slow nodes further. In this way, SAMR can launch backup tasks for map straggler tasks on reduce slow nodes and launch backup tasks for reduce straggler tasks on map slow nodes.
- Experimental result shows that SAMR scheduler can achieve a performance gain up to 37% for MapReduce jobs.

References

1. A. Aboulnaga, Z. Wang, and Z.Y. Zhang. Packing the most onto your cloud. In *Proceeding of the first international workshop on Cloud data management*, pages 25–28. ACM, 2009.
2. F. Ahmad, S. T. Chakradhar, A. Raghunathan, and T. N. Vijaykumar. Tarazu: Optimizing mapreduce on heterogeneous clusters. In *Proceedings of the Seventeenth International Conference on Architectural Support for Programming Languages and Operating Systems*, ASPLOS XVII, pages 61–74, New York, NY, USA, 2012. ACM.
3. L.A. Barroso, J. Dean, and U. Holzle. Web search for a planet: The Google cluster architecture. *IEEE Micro*, 23(2):22–28, 2003.
4. H. S. Bhosale and D. P. Gadekar. Big data processing using hadoop: Survey on scheduling. *International Journal of Science and Research (IJSR)*, 3(10):272–277, 2014.
5. R. Buyya, C.S. Yeo, S. Venugopal, J. Broberg, and I. Brandic. Cloud computing and emerging IT platforms: vision, hype, and reality for delivering computing as the 5th utility. *Future Generation Computer Systems*, 25(6):599–616, 2009.
6. F. Chang, J. Dean, S. Ghemawat, W.C. Hsieh, D.A. Wallach, M. Burrows, T. Chandra, A. Fikes, and R.E. Gruber. Bigtable: A distributed storage system for structured data. In *Proceedings of the 7th USENIX Symposium on Operating Systems Design and Implementation (OSDI 2006)*, 2006.
7. J. Chauhan, D. Makaroff, and W. Grassmann. The impact of capacity scheduler configuration settings on mapreduce jobs. In *Cloud and Green Computing (CGC), 2012 Second International Conference on*, pages 667–674. IEEE, 2012.
8. R. Chen, H. Chen, and B. Zang. Tiled-MapReduce: optimizing resource usages of data-parallel applications on multicore with tiling. In *Proceedings of the 19th international conference on Parallel architectures and compilation techniques*, pages 523–534. ACM, 2010.
9. Q. Chen, M. Guo, Q. Deng, L. Zheng, S. Guo, and Y. Shen. HAT: history-based auto-tuning MapReduce in heterogeneous environments. *The Journal of Supercomputing*, 64(3):1038–1054, 2013.
10. Colby Ranger, Ramanan Raghuraman, Arun Penmetsa, Gary Bradski, and Christos Kozyrakis. Evaluating mapreduce for multi-core and multiprocessor systems. In *HPCA 2007: Proceedings of the 2007 IEEE 13th International Symposium on High Performance Computer Architecture*, pages 13–24, Washington, DC, USA, 2007. IEEE Computer Society.
11. M. De Kruijf and K. Sankaralingam. MapReduce for the cell broadband engine architecture. *IBM Journal of Research and Development*, 53(5):10, 2010.
12. J. Dean and S. Ghemawat. MapReduce: a flexible data processing tool. *Communications of the ACM*, 53(1):72–77, 2010.
13. P. Elespuru, S. Shakya, and S. Mishra. Mapreduce system over heterogeneous mobile devices. *Software Technologies for Embedded and Ubiquitous Systems*, pages 168–179, 2009.
14. W. Fang, B. He, Q. Luo, and N.K. Govindaraju. Mars: Accelerating MapReduce with Graphics Processors. *IEEE Transactions on Parallel and Distributed Systems*, 2010.

15. M.J. Fischer, X. Su, and Y. Yin. Assigning tasks for efficiency in Hadoop. In *Proceedings of the 22nd ACM symposium on Parallelism in algorithms and architectures*, pages 30–39. ACM, 2010.
16. Hadoop. Hadoop home page. http://hadoop.apache.org/, 2011.
17. Jeffrey Dean and Sanjay Ghemawat. Mapreduce: simplied data processing on large clusters. In *OSDI 2004: Proceedings of 6th Symposium on Operating System Design and Implemention*, pages 137–150, New York, 2004. ACM Press.
18. W. Jiang, V.T. Ravi, and G. Agrawal. A Map-Reduce System with an Alternate API for Multi-core Environments. In *2010 10th IEEE/ACM International Conference on Cluster, Cloud and Grid Computing*, pages 84–93. IEEE, 2010.
19. Matei Zaharia, Andy Konwinski, Anthony D. Joseph, Randy Katz, and Ion Stoica. Improving mapreduce performance in heterogeneous environments. In *8th Usenix Symposium on Operating Systems Design and Implementation*, pages 29–42, New York, 2008. ACM Press.
20. Matei Zaharia, Dhruba Borthakur, Joydeep Sen Sarma, Khaled Elmeleegy, Scott Shenker, and Ion Stoica. Job scheduling for multi-user mapreduce clusters. Technical Report UCB/EECS-2009-55, EECS Department, University of California, Berkeley, Apr 2009.
21. K. Morton, M. Balazinska, and D. Grossman. ParaTimer: a progress indicator for MapReduce DAGs. In *Proceedings of the 2010 international conference on Management of data*, pages 507–518. ACM, 2010.
22. J. Polo, D. Carrera, Y. Becerra, J. Torres, E. Ayguadé, M. Steinder, and I. Whalley. Performance Management of Accelerated MapReduce Workloads in Heterogeneous Clusters. In *39th International Conference on Parallel Processing (ICPP2010). San Diego, CA, USA*, 2010.
23. M.M. Rafique, B. Rose, A.R. Butt, and D.S. Nikolopoulos. CellMR: A framework for supporting mapreduce on asymmetric cell-based clusters. In *Parallel & Distributed Processing, 2009. IPDPS 2009. IEEE International Symposium on*, pages 1–12. IEEE, 2009.
24. T. Sandholm and K. Lai. Dynamic proportional share scheduling in hadoop. In *Job Scheduling Strategies for Parallel Processing*, pages 110–131. Springer, 2010.
25. Sanjay Ghemawat, Howard Gobioff, and Shun-Tak Leung. The google file system. In *SOSP 2003: Proceedings of the 9th ACM Symposium on Operating Systems Principles*, pages 29–43, New York, NY, USA, 2003. ACM.
26. M.C. Schatz. CloudBurst: highly sensitive read mapping with MapReduce. *Bioinformatics*, 25(11):1363, 2009.
27. Y. Shan, B. Wang, J. Yan, Y. Wang, N. Xu, and H. Yang. FPMR: MapReduce framework on FPGA. In *Proceedings of the 18th annual ACM/SIGDA international symposium on Field programmable gate arrays*, pages 93–102. ACM, 2010.
28. C. Tian, H. Zhou, Y. He, and L. Zha. A dynamic MapReduce scheduler for heterogeneous workloads. In *Proceedings of the 2009 Eighth International Conference on Grid and Cooperative Computing-Volume 00*, pages 218–224. IEEE Computer Society, 2009.
29. L.M. Vaquero, L. Rodero-Merino, J. Caceres, and M. Lindner. A break in the clouds: towards a cloud definition. *ACM SIGCOMM Computer Communication Review*, 39(1):50–55, 2008.
30. J. Varia. Cloud architectures. *White Paper of Amazon*, http://jineshvaria.s3.amazonaws.com/public/cloudarchitectures-varia.pdf, 2008.
31. R.M. Yoo, A. Romano, and C. Kozyrakis. Phoenix rebirth: Scalable MapReduce on a large-scale shared-memory system. In *Workload Characterization, 2009. IISWC 2009. IEEE International Symposium on*, pages 198–207. IEEE, 2009.
32. M. Zaharia, D. Borthakur, J.S. Sarma, K. Elmeleegy, S. Shenker, and I. Stoica. Job scheduling for multi-user mapreduce clusters. Technical report, Technical Report UCB/EECS-2009-55, University of California at Berkeley, 2009.
33. M. Zaharia, D. Borthakur, J. Sen Sarma, K. Elmeleegy, S. Shenker, and I. Stoica. Delay scheduling: A simple technique for achieving locality and fairness in cluster scheduling. In *Proceedings of the 5th European conference on Computer systems*, pages 265–278, Paris, France, 2010. ACM.

Chapter 8
QoS-Aware Task Reordering for Accelerators

Abstract Modern computers are being outfitted with non-preemptive accelerators, such as GPU and FPGA, to provide the significant compute ability required by emerging large-scale workloads. While an accelerator is able to host multiple applications concurrently, once the tasks are launched to an accelerator, there is no open interface to schedule them. Lacking of task scheduling mechanism on non-preemptive accelerator limits the applicability of to accelerators in many fields. For instance, while the latency-critical services hosted by accelerators in datacenter have diurnal access pattern, it is not applicable to co-locate other applications with the services for improving accelerator utilization. This is because interference when co-locating applications on non-preemptive accelerators may result in the QoS violation of latency-critical applications. Lacking the ability of scheduling tasks on accelerators makes it hard to control the resource allocation between applications. To this end, in this chapter, we present a task scheduling mechanism, Baymax, on non-preemptive processors. After that, as a case study, we use Baymax to improve the accelerator utilization while guaranteeing the Quality-of-Service of latency-critical applications. Using DjiNN, a deep neural network service, Sirius, an end-to-end IPA workload, and traditional applications on a Nvidia K40 GPU, our evaluation shows that Baymax improves the accelerator utilization by 91.3% while achieving the desired 99%-ile latency target for latency critical applications. In fact, Baymax reduces the 99%-ile latency of latency-critical applications by up to 195x over default execution.

8.1 Background and Existing Problems

We refer to accelerators that do not support context switching during kernel execution (such as ASICs, FPGAs and GPUs) as *non-preemptive*. Accelerators are often connected to the host machine through PCIe bus. Before the tasks can be executed,

---

Part of contents in this chapter has been published through International Conference on Architectural Support for Programming Languages and Operating Systems (ASPLOS). Reprinted from Ref. [5], with permission from ACM. Figures 8.1, 8.8, 8.9, 8.10, 8.11, 8.12, 8.13, 8.14, 8.15, and 8.16 in this chapter have been published through International Conference on Architectural Support for Programming Languages and Operating Systems (ASPLOS). Reprinted from Ref. [5], with permission from ACM.

© Springer Nature Singapore Pte Ltd. 2017

Q. Chen and M. Guo, *Task Scheduling for Multi-core and Parallel Architectures*,
https://doi.org/10.1007/978-981-10-6238-4_8

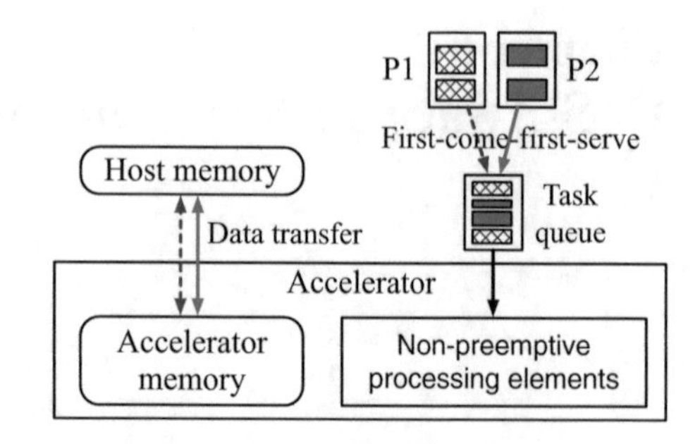

Fig. 8.1 Two applications submit their tasks to the same GPU

the required data need to be first transfer to the accelerator memory through the PCIe bus. Without loss of generality, we use the GPU as our non-preemptive accelerator platform throughout this chapter. Traditionally, an accelerator can only process a single application, and the tasks (aka., kernels) are submitted and executed sequentially. Recently, emerging accelerators, such as GPGPU, already supports multiple applications to submit their kernels concurrently. In this case, the kernels are executed in a *First-Come-First-Serve* manner. Figure 8.1 shows the way that the tasks are executed on a recent GPU when multiple applications use the same GPU.

As shown in the figure, when two applications submit their tasks to the same GPU, the computational tasks (kernels) are sorted according to their arrive time and then executed sequentially. Meanwhile, the data transfer tasks contend for the PCIe bandwidth to transfer their data between host memory and accelerator memory. However, besides the first-come-first-server execution manner, there is no open interface to schedule the kernels in another manner. Lacking the ability to schedule the kernels issued to the accelerator could result in severe problem in real-world scenario. For instance, accelerators have been shown to be particularly suitable for emerging datacenter applications from both performance and total cost of ownership (TCO) perspectives [18]. Therefore, to satisfy the ever-growing user demand at a low cost, datacenters have recently adopted accelerator-outfitted servers for these applications [2, 22]. Meanwhile, since these services generally experience diurnal pattern [17, 21] (leaving the accelerator resources under-utilized for most of the time except peak hours), it is more cost efficient to co-locate latency critical applications and batch applications on accelerators. However, accelerator sharing introduces varying amount of performance interference between co-located applications. Lacking of the ability to schedule accelerator tasks, it is challenging to guarantee that latency critical applications can meet their quality of service targets. In this chapter, we seek to solve the task scheduling problem on non-preemptive accelerators.

8.2 Prior Work on Handling Accelerator Co-location

Prior work [12, 13, 24, 29] has proposed techniques to improve the performance of traditional realtime GPU tasks (e.g., frames per second for video processing) when they are co-located with other GPU tasks. For instance, TimeGraph [29] can be implemented in real-system device driver to re-schedule GPU kernels; SMK [34] can improve the overall system throughput by allowing warps from multiple kernels on the same SMs. GPU-EvR [14] can minimize QoS violation of high priority applications by allocating them more SMs. However, SMK and GPU-EvR require hardware modification thus are not able to be applied in real-system accelerator.

8.2.1 TimeGraph

Kato et al. [29] proposed TimeGraph, a real-time GPU scheduler at the device-driver level for protecting important GPU workloads from performance interference. TimeGraph adopts an event-driven model that synchronizes the GPU with the CPU to monitor GPU commands issued from the user space and control GPU resource usage in a responsive manner.

8.2.1.1 Scheduling Policies

TimeGraph supports two scheduling policies: *Predictable-Response-Time* (PRT) and *High-Throughput* (HT). PRT tries best to guarantee the QoS of high priority applications and HT aims to maximize the overall throughput without considering the priorities of the co-located applications.

PRT Scheduling: In the PRT scheduling policy, when a kernel is scheduled to run, it has to wait for the completion of the preceding GPU kernels. In more detail, when a new GPU kernel arrives at the device driver of a GPU, the kernel is submitted to the GPU immediately if the GPU is currently free. Otherwise, if the GPU is currently busy, the newly arrived kernel much sleep in the wait queue. The kernel with the highest priority in the wait queue is woken up first, once the GPU completes the current kernel. One weakness of this policy is that it may incur QoS violation of high priority applications if the current running low priority kernel long. Furthermore, because PRT needs to make a scheduling decision at every kernel boundary, it incurs overhead that might hurt the overall throughput.

HT Scheduling: In order to reduce the scheduling overhead, the HT policy allows kernels to be submitted to the GPU immediately, if (i) the currently-executing kernel was submitted by the same task, and (ii) no higher-priority tasks are ready in the wait queue. Otherwise, they must suspend in the same manner as the PRT policy. Upon an interrupt, the highest-priority task in the wait queue is waken up, only when the GPU is idle. Compared with PRT policy, the QoS violation could be worse in HT policy because a high priority kernel could be blocked by a great many continuous low priority kernels.

The above two policies are not able to guarantee the QoS of high priority applications but tries to minimize the QoS violation.

8.2.1.2 Resource Reservation Policies

TimeGraph supports two GPU reservation policies: *Posterior Enforcement* (PE) and *Apriori Enforcement* (AE).PE enforces GPU resource usage after the current running warps completes without sacrificing throughput and AE enforces GPU resource usage before GPU warps are submitted using prediction of GPU execution costs at the expense of additional overhead.

The two resource reservation policies are proposed to regulate GPU resource usage for tasks scheduled under the PRT policy. In TimeGraph, each application is assigned a *reserve* that is represented by capacity C and period T. Budget e is the amount of time that an application is entitled for execution. Specifically, the budget is decreased by the amount of time consumed on the GPU, and is replenished by at most capacity C once every period T.

PE Reservation: In the PE policy, if the budget of an application is larger than zero, the application's kernels is allowed to be submitted to GPU. Otherwise, its kernels goes to sleep until the budget is replenished. The budget can be negative, when the task overruns out of reservation. The overrun penalty is, however, imposed on the next budget replenishment. The budget for the next period is therefore given by $e = min(C, e + C)$.

AE Reservation: In AE policy, when a kernel of a application is submitted to TimeGraph, the AE policy first predicts its cost x on the GPU. The kernel can be actually submitted to the GPU, only if the predicted cost x is no greater than the remaining budget of the application. Otherwise, the task goes to sleep until the budget is replenished. The next replenishment amount depends on the predicted cost x and the currently-remaining budget e. If the predicted cost x is no greater than the capacity C, the budget for the next period is bounded by $e = C$ to avoid transient overload. Else, it is set to $e = minx, e + C$. The application can be waken up only when $e \geq x$.

After looking into the above description, we can find that TimeGraph generally relies on priority-based policy to manage GPU kernels. Later in this chapter, we will evaluate the performance of priority-based policy on guaranteeing the QoS of latency-sensitive applications. Similar to TimeGraph, Elliott et al. [12] proposed GPUSync, which also schedule GPU kernels using priority-based policies.

These techniques rely on users to provide task arrival rate, length of time window and the expected GPU time for each type of GPU tasks. Such information is often unavailable in real datacenter environment. In addition, these techniques focus on increasing throughput for high priority tasks, overlooking the long tail latency problem, which is more critical for latency critical applications.

8.2.2 GPU-EvR

Lee et al. [14] proposed GPU-EvR that maps concurrent applications to different streaming multiprocessors (SMs) on the same GPU. These techniques assign a fix proportion of GPU time to high priority tasks but cannot guarantee that the realtime tasks do not violate the QoS requirement [14].

While an active kernel occupies all the SMs in TimeGraph, the SMs can be allocated to multiple concurrent kernels in GPU-EvR. The SMs allocation requires hardware modification and is not applicable directly in emerging real system GPU. GPU-EvR consists of a *workload manager* and a *GPU manager*.

8.2.2.1 Workload Manager

When an application is submitted, its response deadline, profiling data and priority are also submitted to GPU-EvR at the same time. If there are no waiting applications and the GPU has available resources, the application is directly submitted to the GPU. Otherwise, the workload manager classifies the application based on its priority and pushes it into a corresponding queue. Whenever there are available resources in the GPU, the workload manager selects an application from any non-empty waiting queues based on the priority.

Furthermore, in order to present the starvation of low priority applications, the workload manager creates a special application queue, *urgent queue*. At runtime, the workload manager tracks the system time $T_{current}$ and the required response time of each application. Equation 8.1 calculates the time that application A_i has to start to run in order to return before the deadline. In the equation, T_i^{resp} and $E(A_i)$ are the response deadline of A_i and the processing time of A_i, respectively.

$$T_i^{margin} = T_i^{resp} - E(A_i) \tag{8.1}$$

The workload manager compares $T_{current}$ and T_i^{margin} of low priority applications. If $T_{current}$ is close to T_i^{margin}, the workload manager classifies the application A_i as an urgent application and pushes the application onto the urgent queue. Applications in urgent queues are submitted to GPU before all the other applications.

8.2.2.2 GPU Manager

With the above workload manager, the following two cases may happen.

- High priority applications are not submitted with enough resources: In this case, resource reallocation is required to allocate enough GPU resources to high priority applications.
- The application completes its operation on the GPU: After the application completes its work, GPU resources are released and made available to other applications. Therefore, currently executing applications are able to use more GPU resources through resource reallocation.

In more detail, when GPU-EvR starts to reallocate resources, the GPU manager creates a resource reallocation list and obtains the GPU resources for resource reallocation. While creating the resource reallocation list, the GPU manager checks the resource status of the application in a priority order. If the application has enough GPU resources to meet the timing requirement, the GPU manager keeps the current status. However, if the application does not have enough GPU resources, all the lower priority applications are included on the resource reallocation list.

Note that, the GPU manager tries to reallocate resources for applications in a priority order. For an application, if the required additional resources is less than the re-allocatable resources, the GPU manager assigns required resources and updates re-allocatable resources. Otherwise, current re-allocatable resources are assigned to the current application.

Similar to GPU-EvR, Aguilera et al. [24] proposed a technique to guarantee QoS of high priority tasks by spatially allocating them more SMs on a GPU. The two systems assume that programmers can decide how to allocate SMs to the co-located applications. However, commodity GPUs do not support allocating a set of SMs to a specific application.

8.2.3 Simultaneous Multi-kernel (SMK)

Wang et al. [34] introduced a new notion of sharing a GPU that significantly improves resource utilization (both static and dynamic) to boost overall system throughput. Simultaneous Multikernel (SMK) draws an analogy from simultaneous multithreading for CPUs, to increase thread-level parallelism (TLP) of a GPU. SMK exploits kernel heterogeneity to allow fine-grain sharing by multiple kernels within each SM. The fundamental principle is to co-execute kernels with compensating resource usage in the same SM to achieve high utilization and efficiency.

The proposed GPU sharing mechanism aims to co-execute multiple kernels in the same SM, as depicted in Fig. 8.2a. Initially K_0 runs on both SM_0 and SM_1. When K_1 arrives and demands sharing, SMK will let K_0 and K_1 co-execute on both SM_0 and SM_1. This is in contrast to Spart [15], where K_0 and K_1 share the GPU through splitting SMs, as depicted in Fig. 8.2b. Both schemes assume a general execution model where applications do not necessarily arrive at the GPU at the same time, so that sharing can happen dynamically and flexibly. To achieve this, the already executing kernel must be preempted to save part of its context, and context of the incoming kernel must be loaded. With Spart, the amount of context swapped is in unit of an entire SM, which is hefty hundreds of kilobytes of memory traffic overhead. SMK, on the contrary, requires swapping of only partial context of an SM, as explained next.

8.2.3.1 Partial Context Switching

When a new kernel *newK* arrives, the current kernel *K* is partially preempted. The context of *K* is saved to memory one TB at a time, until enough resources are

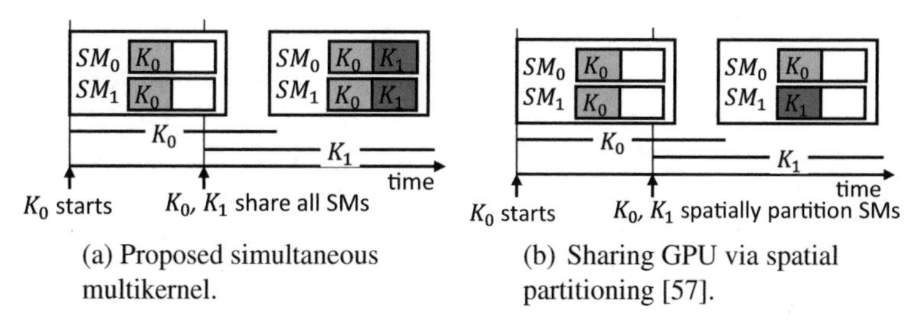

(a) Proposed simultaneous multikernel.

(b) Sharing GPU via spatial partitioning [57].

Fig. 8.2 Sharing GPU enabled by preemption

released to host one TB of $newK$. Hence, Wang et al. [34] terms our preemption mechanism *Partial Context Switching* (PCS). The main distinctive feature of PCS is that preemption takes place without blocking the SM, i.e., the SM continues executing the remaining TBs of K while switching contexts. PCS leads to not only forward progress in kernel execution during a context switch, but also less overhead.

8.2.3.2 Fair Allocation of Static Resources

Unlike Spart where resources are allocated in unit of SMs, SMK makes the SM a divisible resource to better utilized them, improve the overall GPU throughput, and achieve better fairness among sharers. SMK adopts Dominant Resource Fairness (DRF) [11] to define the resource share of a kernel. The intuition behind DRF is that multi-resource allocation should be determined by the maximum share that a kernel requires of any resource. Hence, SMK uses the maximum share of register, threads, shared memory and TB to define the resource share of the kernels (rK) and SMs (rSM).

Wang et al. [34] proposed to allocate the resources for each kernel *before* their TBs are dispatched. Wang et al. [34] termed this strategy as resource partitioning, and the allocated resources for one kernel is called a resource partition. To create resource partitions, the number of TBs of each kernel in each SM is calculated through a iterative procedure, aiming to equalize rK for all kernels on one SM. Then, the TBs are dispatched or swapped according to the generated resource partitions. Our experimental results show that this allocation effectively improves the GPU throughput.

8.2.3.3 Fair Allocation of Dynamic Resource

Wang et al. [34] has observed that kernels may have unfair performance, even though the kernels have a fair share of static resources. Hence, they further develop algorithms to perform dynamic resource allocation, in terms of kernel execution cycles, via warp scheduling.

In SMK, a fair allocation of computing cycles is defined as one where a kernel has a share of cycles ($Quota_k$) that is proportional to the amount required when the kernel is executed exclusively on one SM (x). The proportion is determined by the ratio of resources allocated in SMK (i.e., number of TBs) to resources used in isolated execution (T). SMK obtains x and T by assigning a dedicated SM to each kernel for profiling. During each epoch of execution, each warp scheduler allocates $Quota_k \times Epoch\_length$ number of instructions for kernel k. The quota is decremented when a instruction is issued. If one kernel's quota reaches zero, new instructions of the kernel will be blocked. If all quotas reach zero, new quotas will be calculated and assigned to each kernel.

The proposed allocation of cycles ensures that the number of issued instructions is related to the number of TBs of the kernel present in an SM. As a result, warps of kernels can be relatively fairly scheduled by warp schedulers.

8.2.4 GPU Thread Preemption

At the hardware level, GPU thread preemption [16, 20] is also proposed to intelligently schedule threads for improved hardware utilization. Tanasic et al. [15] proposed a technique that improves performance of high priority processes by enabling preemptive scheduling on GPUs. Wang et al. [33] proposed QoS mechanisms for a fine-grained form of GPU sharing. The QoS support can provide control over the progress of kernels on a per cycle basis and the amount of thread-level parallelism of each kernel. However, the proposed technique requires vendors to add extra hardware extensions and does not work on commodity accelerators. These techniques are out of the scope of this book, thus we do not explain them in detail.

8.3 Real System Investigation on Accelerator Co-location

In this section, through real-system investigation, we show the problem that latency-critical applications suffer from QoS violation at co-location on accelerators. Our real system study uses both latency critical applications and batch applications. Latency critical applications, such as emerging IPA application Sirius [18] and deep neural network service DjiNN [19], run as permanent services on the accelerator, accepting user queries and returning the results with stringent QoS requirement. Batch applications on the other hand do not have QoS requirement but only require high throughput. Both latency critical applications and batch applications consist of various number of tasks (*kernels* and *memcpy tasks*[1]), and the duration of each task also varies across applications. In this experiment, multiple latency critical applications

[1]A task that runs on processing elements is refer as a *kernel* and a task that transfers data through PCI-e bus is refer as a *memcpy task*.

Table 8.1 Benchmarks used in this chapter

Benchmark suite	Workloads
Sirius suite in Sirius [18]	asr, gmm, stemmer (stem)
Tonic suite in DjiNN [19]	dig, face, imc, ner, pos
Rodinia [30]	heartwall (hw), lavaMD (md), cfd, hybridsort (hsort), hotspot (hs), nw, pathfinder (pf)

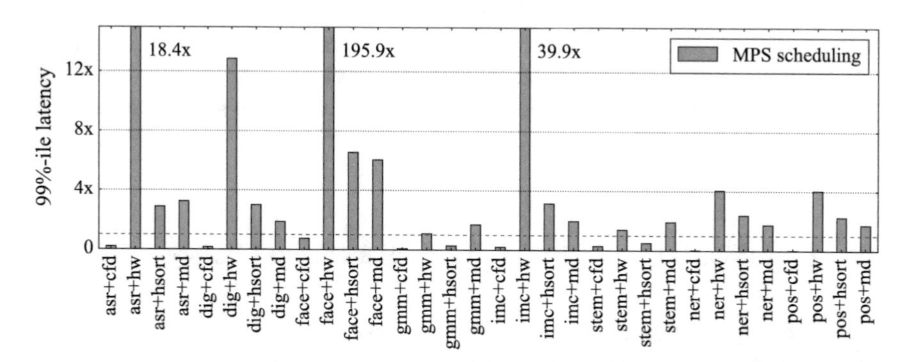

Fig. 8.3 QoS violation of latency critical applications at co-locations with default MPS scheduling policy, when a latency-critical application is co-located with compute-intensive batch applications

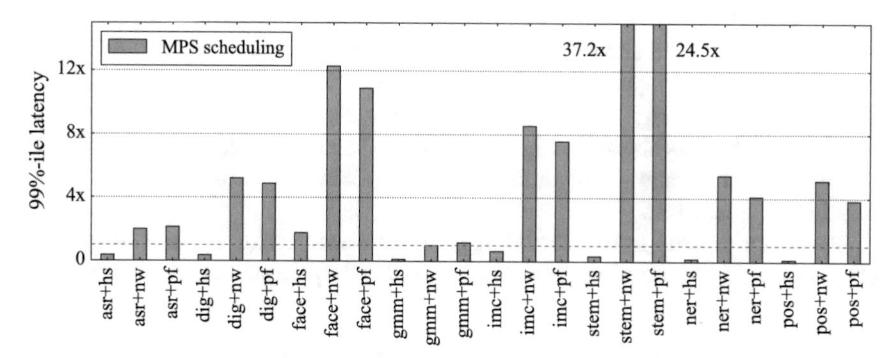

Fig. 8.4 QoS violation of latency critical applications at co-locations with default MPS scheduling policy, when a latency-critical application is co-located with PCIe-intensive batch applications

and batch applications submit kernels and memcpy requests to GPU simultaneously. Table 8.1 lists the used benchmarks throughout this chapter.

Emerging GPU leverages *MPS (Multi-Process Service) scheduling* [23] to enable concurrent sharing of a GPU among multiple applications. Figures 8.3 and 8.4 show the QoS violation when a latency critical application is co-located with compute-intensive/PCIe intensive batch applications on a Nvidia K40 GPU with MPS scheduling policy, respectively. In the two figures, the x-axis indicates the combination of latency critical application and batch application, and the y-axis shows the 99%-ile latency of the latency critical applications normalized to its QoS target (150 milliseconds [17, 32]). The left part of the figure and the right part (shadowed part) of the figure show the results when a latency critical application is co-located with compute intensive batch applications and PCI-e intensive batch applications, respectively. As shown in Fig. 8.3, the 99%-ile latency of latency critical queries in 22 out of the 32 co-locations is much larger than the expected QoS target with default MPS scheduling. As shown in Fig. 8.4, the 99%-ile latency of latency critical queries in 16 out of the 24 co-locations is much larger than the expected QoS target with default MPS scheduling. The 99%-ile latency of latency critical applications is 10.8x of the QoS target on average and up to 195.9x in the worst case. Observed from the above experiment, we can find that it could be problematic to rely on accelerator itself to schedule the tasks.

8.4 Investigation on Priority-Based Scheduling Policy

In order to improve schedule kernels on demand, researchers proposed *priority-based scheduling policy* [12, 29]. In priority-based scheduling policy, each application is given a priority, and the kernels are scheduled according to their priorities. In another word, an accelerator executes high priority kernels first if multiple kernels are ready to run.

Figures 8.5 and 8.6 show the QoS violation when a latency critical application is co-located with compute-intensive/PCIe intensive batch applications on a Nvidia K40 GPU with priority-based scheduling policy, respectively. Adopting priority-based scheduling, as shown in Figs. 8.5 and 8.6, latency critical applications in 33 out of the 88 co-locations still suffer from QoS violation by 1.6x on average (up to 5.2x in the worst case).

The reason priority-based scheduling polices are not capable to guarantee the QoS of latency critical applications (high priority) is that they are not aware of the duration

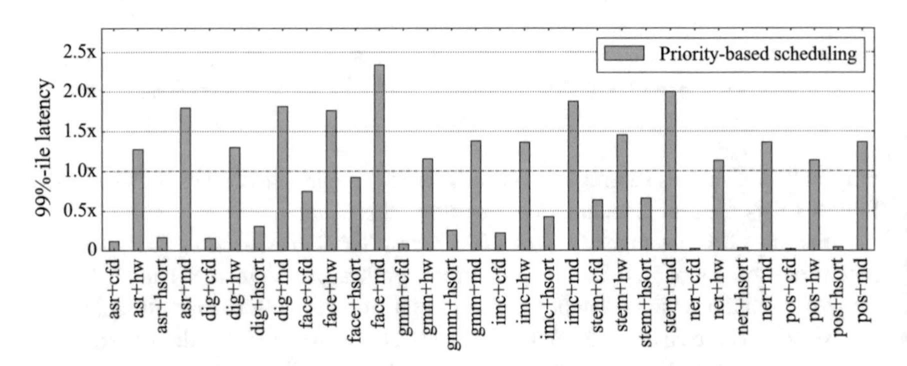

Fig. 8.5 QoS violation of latency critical applications at co-locations with priority-based scheduling policy, when a latency-critical application is co-located with compute-intensive batch applications

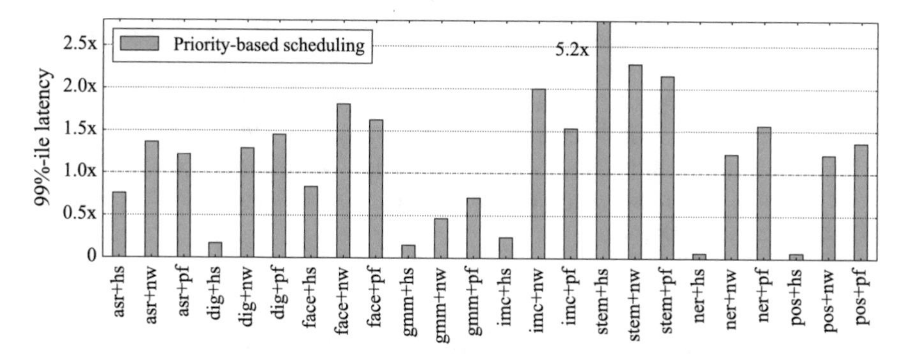

Fig. 8.6 QoS violation of latency critical applications at co-locations with priority-based scheduling policy, when a latency-critical application is co-located with PCIe-intensive batch applications

of tasks. Whenever a latency critical application is not submitting kernels to GPU due to stalls such as CPU synchronization, kernels of batch applications may take over the GPU resource with long duration and high occupancy. Because emerging accelerators (e.g., GPU) are non-preemptive, even if a latency critical kernel becomes ready right after the submission of the long batch kernel, the latency critical kernel would not be executed until the previous kernel completes. In this case, long queuing delay is added to the latency critical kernel, risking QoS violations.

8.5 Design of Task Scheduling Mechanism on Accelerators

As we mentioned before, limited by the existing GPU design, there is no open interface to schedule tasks that are already launched to the GPU. We therefore design a mechanism to schedule tasks on the CPU side. Figure 8.7 presents our design of task scheduling mechanism on accelerators.

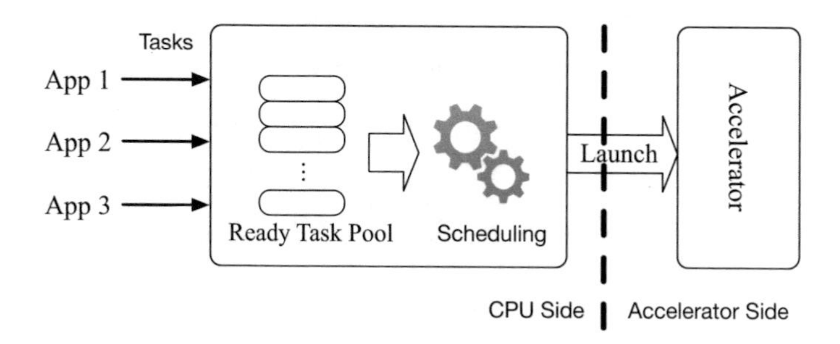

Fig. 8.7 Design of task scheduling mechanism on accelerators

As shown in Fig. 8.7, in the proposed scheduling mechanism, all the tasks submitted to the accelerator are first pushed into a ready task pool managed by the scheduler on the CPU side. This is achieved by simple automatic instrumentation of the original task submission code. The task submission rerouting APIs can be provided to programmers to submit tasks through the scheduler. When a task is pushed into the ready task pool, users can design and implement various scheduling policies to fulfill their specific requirements.

8.6 Case Study: QoS-Aware Task Scheduling on Accelerator

Adopting the design presented above, as a case study, we propose a QoS-aware task scheduler for accelerator, Baymax, to improve the accelerator utilization while guaranteeing the QoS of latency-critical applications.

8.6.1 Root Causes of Long Tail Latency at Co-location

As presented in Sect. 8.4, previous scheduling policies result in the long tail latency of latency-critical applications on non-preemptive accelerators. Before designing the task scheduling policy for guaranteeing the QoS of latency-critical applications, we first explore what are the root causes of long tail latency on non-preemptive accelerators with emerging task scheduling policies.

Figure 8.8 presents two task execution timelines captured with *nvprof* [26], the profiler provided by Nvidia officially, when co-locating *face* (latency critical) and

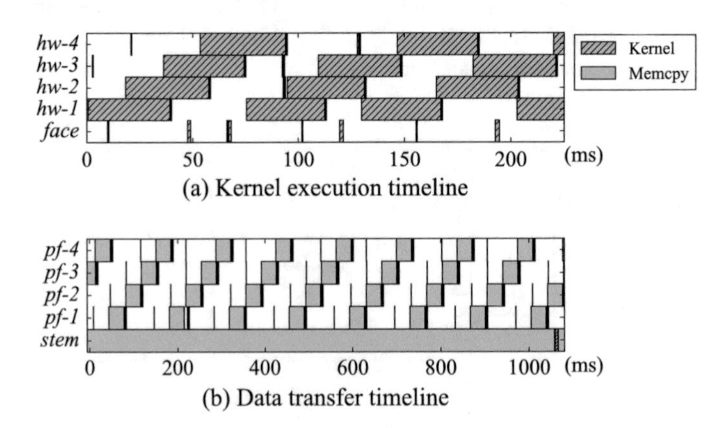

Fig. 8.8 Kernel execution timeline on GPU and data transfer timeline between host memory and accelerator memory

four compute intensive application *hw*, *stem* (latency critical) and four PCI-e intensive application *pf* (details on benchmarks are shown in Table 8.1). Note that the overlapping of green bars in Fig. 8.8a is not kernel preemption, but concurrent kernel execution when using MPS. From the figure we observe that four factors may impact the tail latency of a latency critical application when it is co-located with other applications.

The duration and occupancy of kernels—If the occupancy of a kernel is high, MPS is not able to overlap the kernel with its neighbor kernels to boost concurrent kernel execution. In this case, if the duration of batch kernels is long, the execution of latency critical kernels will be delayed significantly.

The kernel scheduling order—Accelerators, such as GPUs, schedule kernels in the same order as they arrived (even if neighbor kernels can run concurrently when the kernel occupancy is small). If the co-located batch applications submit kernels frequently, the latency critical application will be delayed by a large amount of batch kernels.

The number of kernels in a latency critical query—The more kernels a latency critical query has, the longer its tail latency could be, because every kernel in the latency critical query can be delayed by batch kernels. For example, as shown in Fig. 8.8a, every kernel of *face* is delayed by at least two kernels of *hw*.

The contention on PCI-e bandwidth—If batch applications consume high PCI-e bandwidth, latency critical applications may suffer from slow data transfer due to the contention on PCI-e bandwidth. For example, as shown in Fig. 8.8b, the memcpy task of *stem* is severely slowed down to more than 1000 milliseconds from only 15 milliseconds when it is running alone. This slow down in turn results in long tail latency.

Based on the identified root causes of long tail latency, to improve the utilization of non-preemptive accelerator while guaranteeing the QoS of latency critical applications, Baymax should have the following four abilities.

- Baymax should be able to predict the duration of each kernel and memcpy task. In this case, Baymax can quantify the impact of each task on the end-to-end latency of latency critical applications.
- Baymax should be able to re-order all the kernels issued to the same accelerator, no matter how they are submitted by the co-located applications.
- For a latency critical query, Baymax should be able to limit the overall time delayed by the co-located applications regardless of the number of kernels in the query.
- Baymax should be able to monitor realtime data transfer pressure on PCI-e bus and mitigate PCI-e bandwidth contention.

8.6.2 Design of Baymax

Following the design guidelines of task scheduling mechanism on accelerators presented in Sect. 8.5, Fig. 8.9 presents the design overview of Baymax.

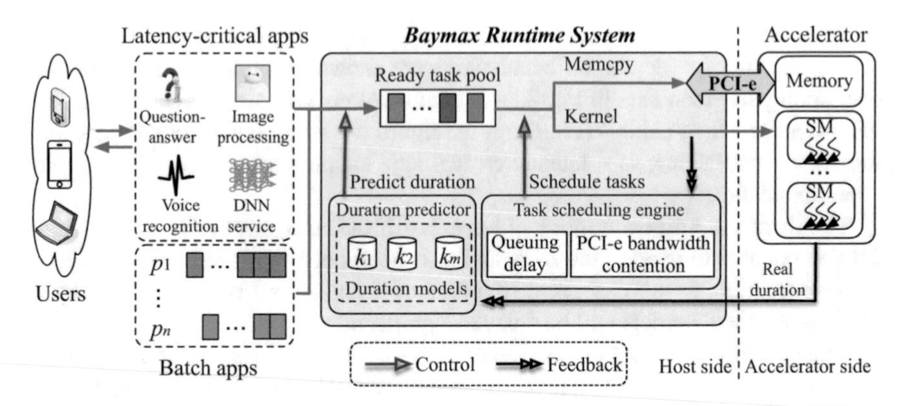

Fig. 8.9 Design of Baymax

In Baymax, all the tasks submitted to the accelerator are first pushed into a ready task pool managed by Baymax on the CPU side. This is achieved by simple automatic instrumentation of the original task submission code. The task submission rerouting APIs can be provided to programmers to submit tasks through Baymax. When a task is pushed into the ready task pool, the task duration predictor first predicts its duration leveraging regression models (Sect. 8.7).

The *task scheduling engine* periodically iterates over all the tasks in the ready task pool and decides whether each task can be launched to GPU. The scheduling decision is based on the QoS target of latency critical applications and the predicted duration of each task. If the task is a kernel and its predicted duration is larger than the realtime QoS headroom of any active latency critical query, the kernel will stay in the ready task pool. Otherwise, the kernel is launched to GPU (Sect. 8.8). On the other hand, if the task is a memcpy task, the engine decides whether to launch the task based on realtime data transfer pressure on PCI-e bus (Sect. 8.9).

8.7 Task Duration Modeling in Baymax

In this section, we present the modeling methodology used by Baymax to predict the duration of GPU tasks.

8.7.1 Task Duration Predictor

Baymax builds duration models for three types of GPU tasks: *memcpy*, *hand-written kernel*, and *library call*. Hand-written kernels are the kernels defined and written by programmers. Besides writing their own hand-written kernels, an application can

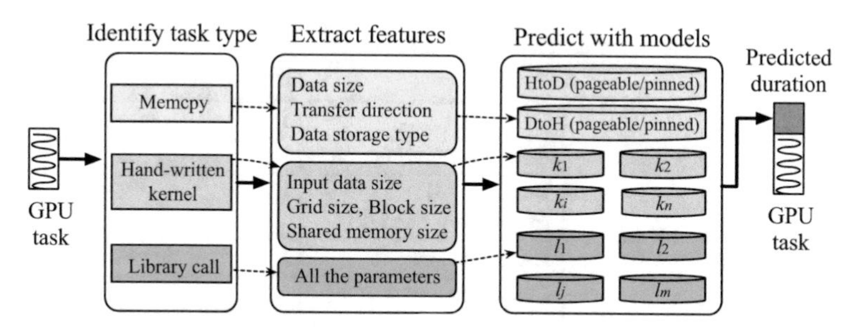

Fig. 8.10 Predict the duration of GPU tasks (memcpy, hand-written kernel, and library call)

also call APIs defined in highly-optimized GPU libraries (e.g., cuDNN [28] and cuBLAS [6]). For the three types of GPU tasks, Fig. 8.10 shows the methodology to predict their duration.

When a task is submitted to Baymax, the duration predictor identifies the type of the task, extracts the representative features, and selects pre-trained duration model according to the name of the GPU task (function name for hand-written kernels and API name for library calls). Once the duration model is found, Baymax predicts the duration of the task using the extracted features and the model, and attaches the predicted duration to the task. After that, the task is pushed into the ready task pool waiting for launching to GPU.

8.7.2 Selecting Representative Features

It is challenging to predict the duration of GPU tasks because there is very limited information we can obtain at runtime. Although *nvprof* [26] provides comprehensive performance metrics after measuring the entire task execution, no performance information can be accessed before the task is executed. It is not applicable to rely on these metrics to predict the duration of GPU tasks.

The only information we can obtain before a GPU task is executed includes its configurations (e.g., grid size, block size etc.) and the parameters passed to the task. We further select the information that strongly impact a task's duration on GPU (e.g., input scale and task configurations) as representative features. Empirically, to capture the correlation between features of a GPU task and its duration, as listed in Table 8.2, we select different features for different types of GPU tasks.

For a memcpy task, we select data size, data transfer direction and data storage type as its representative features. Data to be transferred from/to GPU can be stored in either *pageable memory* or *pinned memory* [8]. It is much faster (around 4x faster) to transfer data from pinned memory compared with pageable memory while more time is needed to initialize pinned memory when it is allocated.

For a hand-written kernel, we use kernel configuration and input data size as the features. The grid size and the block size determine the scale of thread level par-

Table 8.2 Features selected for different types of tasks

Task type	Features	Dimension
Memcpy	Data size	1
	Data transfer direction	1
	Data storage (pageable/pinned [8])	1
Hand-written kernel	Input data size	1
	Grid size ($X \times Y \times Z$)	3
	Block size ($X \times Y \times Z$)	3
	Size of required shared memory	1
Library call	All the parameters	/

allelism on GPU and the GPU occupancy of the kernel, which significantly affect its duration; the size of required shared memory (both *static shared memory* and *dynamic shared memory*) reflects the efficiency of the kernel leveraging the memory hierarchy on GPU. We train different duration models for hand-written kernels executing different functions, because they often have totally different characteristics.

A library call may consist of multiple kernels, while the actual kernels and their configurations are hidden behind the API. Therefore, we treat all the kernels in a library call as a whole, and use all the parameters of the API as its representative features. For several widely used libraries (i.e., cuBLAS and cuDNN), we only need to train models for them once and use the models in all applications.

Besides fine-grained GPU tasks, the duration predictor also predicts the solo-run duration of each latency critical query when the query is first launched. For a latency critical query, we select its input data size as its representative feature.

8.7.3 Low Overhead Prediction Models

The QoS target of a latency critical query is in the granularity of hundreds of milliseconds to support smooth user interaction [17, 32]. Therefore, choosing the modeling techniques with low computation complexity and high prediction accuracy for the online duration predictor becomes critical.

We evaluated a spectrum of widely used prediction models (e.g., *Linear Regression* (LR) [10], *Approximate Nearest Neighbor* (ANN) [3], *K-Nearest Neighbor* (KNN) [9] and *Support Vector Machines* (SVM) [4]) to predict task duration and eventually selected LR and KNN for their high accuracy and low overhead. While LR assumes the linear relationship between input and output variables, KNN regression holds no such assumption. Therefore using both LR and KNN allows us to achieve accurate prediction for both linear and non-linear relations. Other evaluated models either require longer calculation time with no accuracy improvement (e.g.,

SVM), or cannot provide satisfactory accuracy (e.g., ANN). Both KNN and LR achieve low prediction overhead. According to our measurement on real hardware, the duration prediction overhead with KNN model and LR model in Baymax is under 0.05 millisecond.

Suppose a task has p representative features. Let X_i represent an input sample with p features $(x_1, x_2, \ldots, x_p)$, and n represent the total number of input samples $(i = 1, 2, \ldots, n)$. The linear regression model is defined as Eq. 8.2, and the Euclidean distance for KNN model between sample X_i and $X_l (l = 1, 2, \ldots, n)$ is defined as Eq. 8.3. In our case, the input is the task features and the output is the predicted task duration. The primary computation of KNN is to calculate the Euclidean distance between the predicting and training samples, which can be accelerated with different tree searching algorithms such as K-D tree and ball tree. We pick the most efficient KNN searching algorithm when training prediction model according to the number of samples and the number of features in every sample.

$$y_i = \beta_1 x_{i1} + \ldots + \beta_p x_{ip} + \varepsilon_i, i = 1, \ldots, n \tag{8.2}$$

$$d(X_i, X_l) = \sqrt{(x_{i1} - x_{l1})^2 + \ldots + (x_{ip} - x_{lp})^2} \tag{8.3}$$

8.7.4 Minimizing Prediction Error

To achieve high prediction accuracy, we apply both KNN and LR to each task in both latency critical and batch applications, and choose the model that fits the data most to predict the task duration at runtime. As shown in Sect. 8.7.5, LR model and KNN model achieve different prediction accuracy for latency critical applications and batch applications respectively. Since the duration models are trained offline with the profiled performance samples from the workloads, more sample data is usually effective to improve the accuracy of the duration models. Especially, in WSCs, the workloads become stable after certain time scale and the models become more accurate with periodical updates. Moreover, the duration predictor detects the prediction deviation at runtime. If the deviation exceeds a certain threshold, incremental update [7] and parallel update [31] can be applied during runtime with low overhead to refine the duration models, which continuously improves the accuracy of the duration prediction.

8.7.5 Prediction Accuracy

In this section, we present the accuracy of the task duration predictor in Baymax. The representative features for different types of tasks are listed in Table 8.2, and the benchmarks can be found in Table 8.1. The prediction error for the duration of task

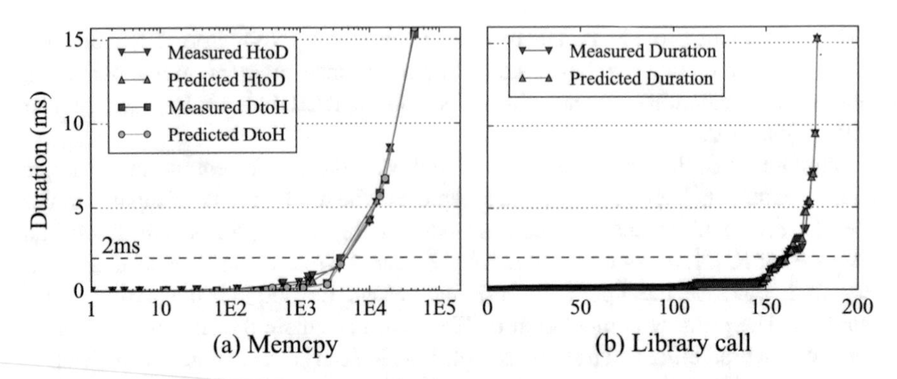

Fig. 8.11 Prediction error for the duration of memcpy tasks and library calls. In **a**, the x-axis is the size of data to be transferred (KB); In **b**, the x-axis is the library calls. Baymax achieves 3.2 and 6.2% prediction errors on average for memcpy and library call respectively

t (memcpy, hand-written kernel or library call) is calculated in Eq. 8.4.

$$Err_t = \frac{\left|Duration_t^{predicted} - Duration_t^{measured}\right|}{Duration_t^{measured}} \tag{8.4}$$

To construct the training and testing data sets for our prediction models, we collect a large amount of samples, and randomly choose 90% of the samples to train the model and use the rest to test. For KNN model, we choose the number of nearest neighbors to be 5 ($K = 5$).

8.7.5.1 Prediction Accuracy for Memcpy

In order to build duration models for memcpy tasks, we create a micro kernel to transfer data between main memory and GPU global memory with arbitrary input sizes. The range of data transfer size in our experiment reflects the actual size of memcpy tasks cross all the benchmarks. As shown in Fig. 8.11a, with the tested data size profiled from all the benchmarks, LR model is able to accurately predict the duration of memcpy across all workloads, which also in accordance with existing literature. The average prediction error is smaller than 3.2%, when the duration is longer than two milliseconds. Thus, Baymax uses LR to predict the duration of memcpy tasks.

8.7.5.2 Prediction Accuracy for Library Call

Library calls take a large portion of GPU execution time across emerging latency critical applications. All the library APIs used in the benchmarks are listed in Table 8.3.

Table 8.3 Frequently used library APIs

Library	API name
cuBLAS [6]	sgemm/dgemm
cuDNN [28]	convolutionForward, addtensor4d poolingForward, activationForward, softmaxForward

These library calls control which kernel to launch as well as the launch configuration with detailed information hidden behind the APIs.

To build duration model for a library call, we analyze every parameter to the library call according to its API definition and extrapolates the size of the input based on the number as well as the data type of the input parameters. Using the input size as the representative feature available at runtime, the prediction fits well into linear regression model as shown in Fig. 8.11b, which is consistent with the findings in prior work [27, 28]. Across all the 180 calls of the library APIs in all the benchmarks, our models can precisely predict the duration of library calls with the prediction error smaller than 6.2%, when the duration is longer than two milliseconds.

If the duration of a library call or a memcpy task is shorter than two milliseconds, even if its duration is not predicted precisely, it will not affect the latency of the co-located applications seriously.

8.7.5.3 Prediction Accuracy for Hand-Written Kernel

The behaviors of hand-written kernels are quite diverse across benchmark suites. While Rodinia is composed of classic HPC workloads that exhibit high thread level divergence on GPU, workloads in Sirius and Tonic are speech recognition, nature language processing and DNN computation that rely on large matrix multiplication with almost no divergence. To build duration models for hand-written kernels, we collect performance samples, including features and duration, using *nvprof* [26]. Note that most of the workloads in Rodinia contain iterative kernel invocations in their implementations and we treat each kernel invocation as an individual sample. To provide rigid validation, we use different samples to train model and to evaluate prediction accuracy.

As shown in Fig. 8.12, no single regression model fits both latency critical and batch applications perfectly. In general, KNN works better than LR for Rodinia since in some cases (e.g., *hs* and *md*) the prediction of LR goes extremely wrong. This observation reveals that the duration of a kernel and its inputs do not always have a linear relationship. Whereas for Tonic suite and Sirius suite, the computation is more regular and predictable, LR has more advantage over KNN with a constrained sample dataset. The average prediction error of KNN for the kernels in Rodinia is 7.2% on average, and the prediction error of LR for Sirius suite and Tonic suite is 5.8% on average.

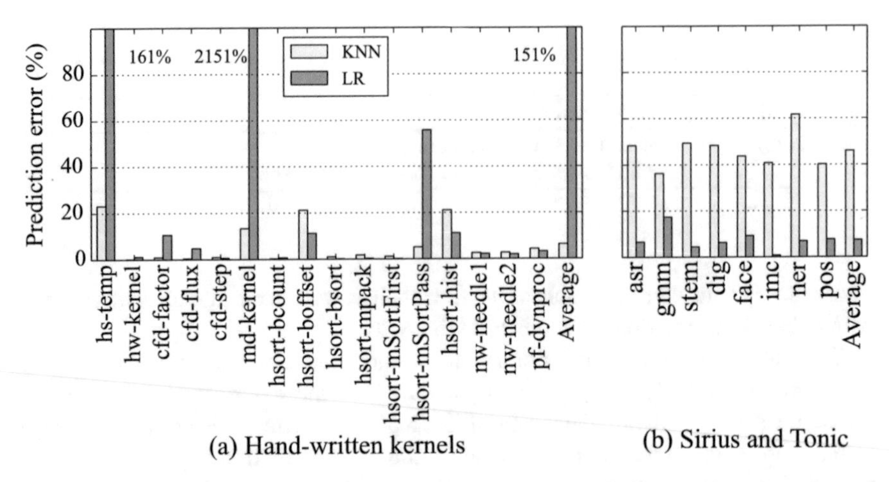

Fig. 8.12 Prediction error for the duration of Sirius, Tonic, and hand-written kernels in Rodinia. KNN model achieves 7.2% prediction error for Rodinia; LR model achieves 5.8% prediction error for Sirius suite and Tonic suite

8.8 Scheduling Hand-Written Kernels and Library Calls

In this section, we describe the policy used to schedule hand-written kernels and library calls in Baymax. For ease of description, a kernel can be either a hand-written kernel or a library call in this section.

8.8.1 Breaking down the End-to-end Latency

It is important to understand the end-to-end latency breakdown of a latency critical query when it is co-located with other applications before diving into the QoS-aware task scheduling policy. We first assume the co-located applications do not contend for PCI-e bandwidth. We will discuss the scheduling policy for tasks that transfer data through PCIe bus in Sect. 8.9.

Figure 8.13 presents the end-to-end latency breakdown of a latency critical query Q when it is co-located with other applications. The end-to-end latency of a query is the time from the first kernel of the query is issued to the last kernel of the query is returned. As shown in the figure, Q's end-to-end latency is composed of three parts. The first part is the processing time of the queued kernels (black kernels in Fig. 8.14) that are issued before k_1 gets executed (denoted by T_q). The second part is the processing time of Q's own kernels (denoted T_{self}). The last part is the processing time of the kernels (line-filled and white kernels) from the co-located applications between k_1 and k_n (denoted by T_{other}).

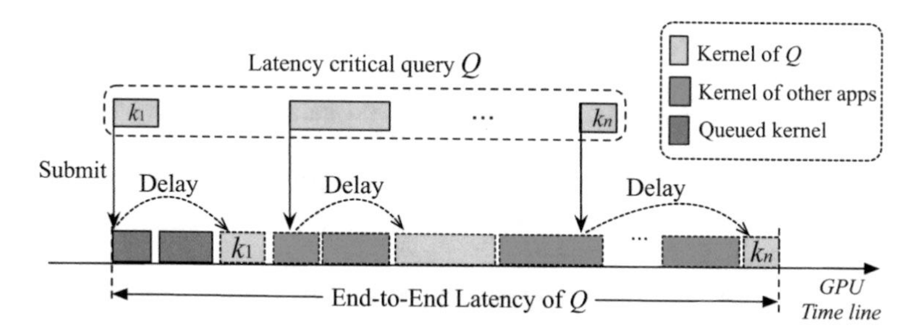

Fig. 8.13 End-to-end latency breakdown of a latency critical query Q when it is co-located with other applications

8.8.2 Scheduling Policy

Based on the end-to-end latency breakdown and the precisely predicted duration of tasks, we can schedule computational tasks carefully so that the end-to-end latency of a query always smaller than the QoS target. In this case, the utilization of hardware can be maximized while still guaranteeing the QoS of latency critical applications. Without loss of generality, let T_{tgt} represent the QoS target of query Q. Only if $T_{self} + T_q + T_{other} \leq T_{tgt}$, Q's QoS is satisfied. In the equation, T_{self} is predicted according to the prediction model proposed in Sect. 8.7. In this case, to guarantee Q's QoS, the task scheduling engine in Baymax monitors T_q and reduces T_{other} as follows.

8.8.2.1 Monitoring Queued Time

To estimate the queuing delay a latency critical query will experience, T_q, Baymax sums up the predicted duration of all the kernels that are already issued to GPU by the task scheduling engine but are not yet executed (still waiting in the GPU queue). Specifically, once a kernel is issued to GPU, we add its predicted duration to T_q, the duration of all the un-executed kernels on GPU. Once a kernel completes, we subtract its predicted duration from T_q.

To eliminate the situation that a latency critical query is significantly delayed by the queued-up kernels on GPU, even if no active latency critical query is running on the GPU, Baymax makes sure that T_q is smaller than the QoS target T_{tgt}. If $T_q > T_{tgt}$, Baymax would not issue any kernel to GPU until some kernels complete. This method would not reduce the GPU utilization because the kernel will be queued up on GPU even if it is issued to GPU.

8.8.2.2 Calculating QoS Headroom

As discussed above, T_{self} and T_q are known and cannot be reduced when Q is launched. In this case, to guarantee Q's QoS, Baymax makes sure that $T_{other} \leq T_{tgt} - T_{self} - T_q$. We use T_{hr} to represent the free GPU time left for kernels from the co-located applications during the execution of Q (referred as *QoS headroom*). When the first kernel of Q is launched, $T_{hr} = T_{tgt} - T_{self} - T_q$.

Based on T_{hr}, the task scheduling engine periodically iterates over the ready task pool to check whether each kernel can be safely issued to GPU without causing any QoS violation. Suppose the predicted duration of a kernel is t. If t is larger than T_{hr}, the kernel is delayed until Q completes. On the other hand, if t is smaller than T_{hr}, the kernel is launched to GPU, and at the same time, T_{hr} is reduced by t.

8.8.2.3 Dealing with Multiple Active Latency Critical Queries

When multiple latency critical queries are active, more complexity is introduced when calculating the headroom of each latency critical query. Figure 8.14 describes the method to calculate T_{hr} of query Q when multiple latency critical queries are active. As shown in the figure, if query Q_i is still active when the first kernel of query Q is launched, the un-executed kernels of Q_i have to be completed before T_d so that the QoS of Q_i is satisfied. In this case, when we calculate T_{hr} for Q, the GPU time reserved by the un-executed kernels of Q_i need to be subtracted from T_{tgt} as well. Therefore, we monitor the GPU time each active query still needs to complete the whole query. For Q_i in Fig. 8.14, we estimate Q_i's remaining GPU time by subtracting the time of its completed kernels from its estimated overall GPU time (T_{self} of Q_i).

Suppose there are n active latency critical queries when Q is launched. Let t_1, ..., t_n represent the remaining GPU time required by the n active latency critical queries respectively. Equation 8.5 calculates Q's QoS headroom when it is issued.

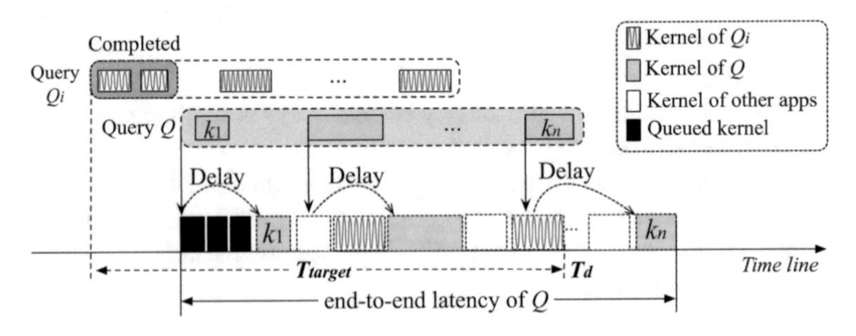

Fig. 8.14 Calculating QoS headroom of Q when its first kernel is launched

$$T_{hr} = T_{tgt} - T_q - T_{self} - \sum_{i=1}^{n} t_i \tag{8.5}$$

When multiple queries are active, if the predicted duration of a kernel (denoted by t) is larger than the QoS headroom of any active query, the kernel will be delayed. Otherwise, the kernel is launched and the QoS headroom of each latency critical query is reduced by t.

It is worth noting that Baymax would not result in starvation of any latency critical query even if multiple queries are active concurrently. latency critical kernels are issued in an FIFO order and a batch kernel can be issued only when it will not result in QoS violation of any active latency critical query.

8.8.2.4 Adapting to Concurrent Kernel Execution

In Sect. 8.8.2, we assume that a GPU is not able to concurrently execute multiple kernels. Actually, leveraging emerging MPS technique [23], a GPU is able to execute multiple independent kernels that have low occupancy concurrently.

When concurrent kernel execution happens, T_{hr} calculated in Eq. 8.5 is smaller than the real GPU time available for the co-located applications. In this case, the GPU utilization is not maximized because there is actually more GPU time can be used to process batch applications while guaranteeing the QoS of all the active latency critical queries.

To further increase GPU utilization when MPS is enabled, as shown in Fig. 8.15, when kernel k_i of Q is submitted to the ready task pool, Baymax updates the QoS headroom of Q. In this way, the time saved from previous concurrent kernel execution can be refilled to the QoS headroom for executing batch applications. Based on Eq. 8.5, the QoS headroom of Q when it submits k_i can be calculated in Eq. 8.6.

$$T_{hr} = (T_{tgt} - T_{used}) - T_q - (T_{self} - \sum_{j=1}^{i} T_j) - \sum_{i=1}^{n} t_i \tag{8.6}$$

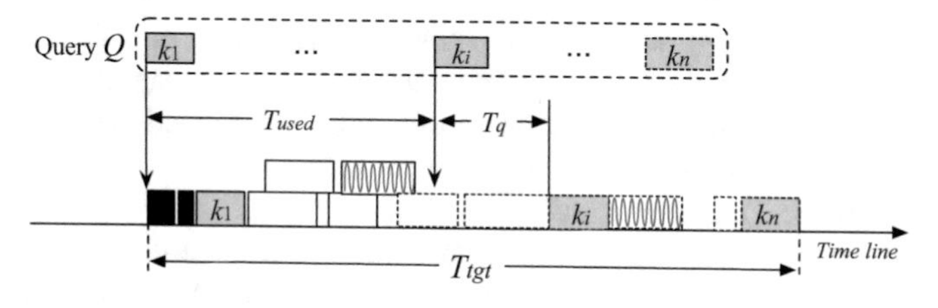

Fig. 8.15 Updating QoS headroom of Q when it submits task k_i, if concurrent kernel execution is enabled

In the equation, T_j is the processing time of kernel k_j, $T_{self} - \sum_{j=1}^{i} T_j$ is the remain GPU time reserved by Q itself, T_{used} is the time from the beginning of Q to k_i is submitted, T_q is the realtime queuing time, t_i is the remaining GPU time required by the active latency critical queries launched before Q as calculated and defined in Sect. 8.8.2.3.

In summary, the QoS headroom of a latency critical query will be updated when a kernel of the co-located applications is launched to GPU and when a new kernel of the query is submitted to the ready task pool.

8.9 Scheduling Data Transfer Tasks

If the co-located applications only have computational tasks (hand-written kernels and library calls), the scheduling proposed in Sect. 8.8 can already guarantee the QoS. However, real-world applications consist of both computational kernels and data transfer tasks. Even if the hand-written kernels/library calls are re-ordered as presented in Sect. 8.8, without considering PCI-e bandwidth contention caused by data transfer tasks, latency critical applications may still suffer from severe QoS violation. In this section, we analyze the impact of PCI-e bandwidth contention on CPU-accelerator data transfer rate per memcpy task and mitigate the contention for achieving QoS of latency critical applications.

8.9.1 Characterizing PCI-e Bandwidth Contention

Remember that data to be transferred from/to GPU can be stored in either *pageable memory* or *pinned memory* [8]. It is much faster (around 4x faster) to transfer data from pinned memory compared with pageable memory while more time is needed to initialize pinned memory when it is allocated. To this end, Fig. 8.16 reports the data transfer rate of a latency critical application *stem* when it is co-located with several applications that transfer data in the same direction. Data transfers in different directions do not interfere with each other, because PCI-e bus supports full-duplex communication. In the figure, the legends show the data transfer direction. For example, "HtoD_pageable_pinned" means *stem* transfers data from pageable memory to GPU, while the co-located applications transfer data from pinned memory to GPU. From the experiment, we have two main observations.

Observation 1: Transferring data from and to pageable memory degrades the performance of its co-located memcpy tasks only when more than three memcpy tasks are running concurrently ("*_*_pageable" in Fig. 8.16). As shown in the figure, when *stem* uses pageable memory and transfers data through PCI-e bus alone, the achieved data transfer rate is 3,150MB/s. Because the theoretical peak bandwidth of 16x PCI-e 3.0 bus used in our platform is 15,800MB/s and the effective bandwidth is 12,160MB/s [1], the bus can only support $\lfloor \frac{12160}{3150} \rfloor = 3$ memcpy tasks to transfer

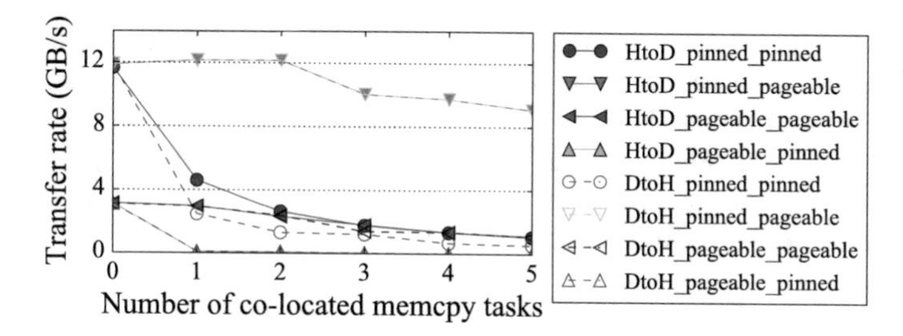

Fig. 8.16 Data transfer rate of *stem* when it is co-located with applications that transfer data in the same direction

data in their full speeds in the same direction. We generalize this observation in Sect. 8.9.2.

Observation 2: A single memcpy task that transfers data from/to pinned memory would severely degrade the performance of its co-located memcpy tasks ("*_*_pinned" in Fig. 8.16). As shown in the figure, transferring data from/to pinned memory requires up to 11,883MB/s PCI-e bandwidth, which saturates the whole PCI-e bus. In this case, all the other memcpy tasks will be queued up and have to wait for its completion.

8.9.2 Scheduling Policy

Baymax mitigates QoS violations due to PCI-e bandwidth contention by reducing the number of concurrent memcpy tasks and considering data transferring delay when calculating QoS headroom for active latency critical queries.

Let BW_{peak} represent the effective PCI-e bandwidth, BW_{memcpy} represent the peak data transfer rate from/to pageable memory per memcpy task. According to observation 1, to make sure that memcpy tasks of latency critical applications can always transfer data in full speed, Eq. 8.7 calculates the number of active batch memcpy tasks N_{tr} that we should allow in each direction. For our platform N_{tr} is two.

$$N_{tr} = \lfloor BW_{peak}/BW_{memcpy} \rfloor - 1 \tag{8.7}$$

Baymax periodically iterates over the ready task pool to check whether each memcpy task can safely start to transfer data. If the memcpy task is from a batch application and there are already N_{tr} active memcpy tasks, the task is delayed until one memcpy task completes. If the memcpy task is from a latency critical query, it is directly issued to GPU to minimize queuing delay.

According to the second observation, if a memcpy task *mc* uses pinned memory, it may severely delay the data transfer of latency critical queries. Let *t* represent the predicted duration of *mc*. If *t* is larger than the QoS headroom of any active latency critical query, *mc* will not be launched. Otherwise, *mc* can start to transfer data, but to avoid QoS violation due to the possible queuing delay caused by *mc*, the QoS headroom of every active latency critical query is reduced by t. This method would not degrade the accelerator utilization. If *mc* does not cause severe queuing delay, the QoS headroom of each active latency critical query will be refilled when a new task is launched as described in Sect. 8.8.2.4.

8.10 Performance of Baymax

In this section, we evaluate whether Baymax that schedules accelerator tasks on CPU side, can successfully guaranteeing the QoS of latency-critical applications while maximizing the accelerator utilization.

8.10.1 Experimental Configuration

We evaluate Baymax using Nvidia GPU K40. Note that Baymax does not rely on any special hardware features or characteristics of K40 and treats it as a generic non-preemptive accelerator. The detailed setups are summarized in Table 8.4. MPS [23] is enabled to allow concurrent kernel execution on GPU. As we already listed in Table 8.1, we use *Tonic suite* [19] in DjiNN and *Sirius suite* [18] in Sirius as the latency critical applications; use eight most compute intensive and three most PCI-e intensive applications from *Rodinia* [30] as batch applications. In order to evaluate the impact of memcpy tasks using both pageable memory and pinned memory, we configure *hs* to use pageable memory, *pf* and *nw* to use pinned memory.

Furthermore, the QoS is defined as the 99%-ile latency, and the accelerator utilization is measured as the ratio of batch application execution time to the whole co-location execution time.

Table 8.4 Hardware and software specifications

	Specifications
Hardware	CPU Intel Xeon E5-2620 @ 2.10GHz Nvidia GPU Tesla K40
Software	CentOS 6.6 x86_64 with kernel 2.6.32-504 CUDA Driver 340.29, CUDA SDK 6.5, CUDA MPS

8.10.2 QoS and Throughput

In this section, we evaluate the effectiveness of Baymax in increasing the accelerator utilization while satisfying the QoS requirement of latency critical applications.

Figure 8.17 presents the average latency, 99%-ile latency of latency critical queries, and the improved accelerator utilization when latency critical applications are co-located with batch applications. In the figure, "Baymax" updates the QoS headroom of each latency critical query when a new kernel is issued to squeeze the extra QoS headroom benefited from concurrent kernel execution as presented in Sect. 8.8.2.4. "Baymax-NC", on the contrary, does not squeeze the extra QoS headroom.

Figure 8.17a and b show that both Baymax-NC and Baymax are able to effectively satisfy the QoS for latency critical applications under different pair-wise co-locations. On the contrary, default MPS scheduling [23] and priority-based scheduling [12, 29] cannot satisfy the QoS for latency critical applications as presented in Sects. 8.3 and

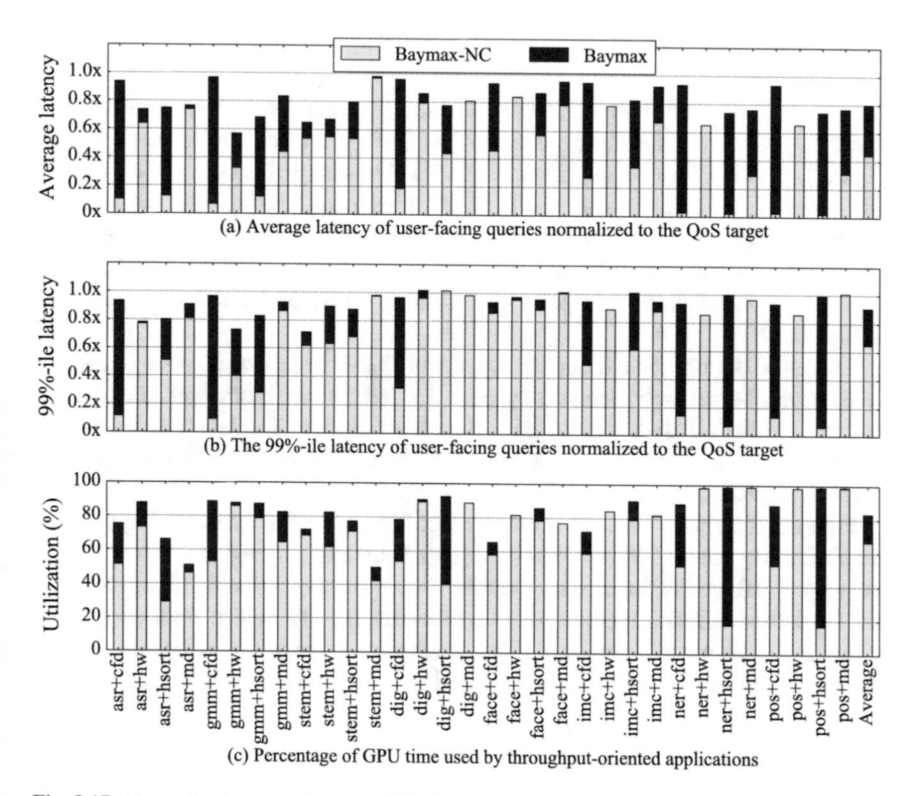

Fig. 8.17 Normalized average latency, 99%-ile latency of latency critical queries, and accelerator utilization when latency critical applications are co-located with compute-intensive batch applications. Different from Baymax, Baymax-NC does not consider concurrent kernel execution

8.4. With MPS scheduling and priority-based scheduling, the 99%-ile latency of latency critical queries is up to 195.9x and 5.2x of the QoS target, respectively.

Figure 8.17a and b also show that the average latency and 99%-ile latency of latency critical queries in Baymax is higher than in Baymax-NC. This is because Baymax squeezes more QoS headroom to trade off higher GPU utilization. As shown in the Fig. 8.17c, Baymax-NC increases the accelerator utilization by 70.8% on average, and Baymax further increases the average accelerator utilization by 11.4%. The reason of utilization increasing is that Baymax can utilize the saved GPU time from concurrent kernel execution to execute more batch kernels.

Observed from Fig. 8.17, for some co-location pairs (e.g., *dig+hsort* and *ner+hw*), the accelerator utilization is not increased using Baymax compared to Baymax-NC. This is because the kernels of these batch applications have large GPU occupancy. In this case, MPS does not have chance to execute multiple kernels concurrently and Baymax cannot squeeze extra GPU time for batch applications.

8.10.3 Scheduling Data Transfer Tasks

As presented in Sect. 8.9, Baymax also mitigates PCI-e bandwidth contention through scheduling data transfer tasks for achieving QoS of latency critical applications. Figure 8.18 shows the average latency and 99%-ile latency of latency critical queries when they are co-located with PCI-e intensive batch applications. As shown in the figure, the QoS requirement of latency critical queries cannot be satisfied if PCI-e bandwidth contention is not mitigated (shown as "Baymax-NP" in Fig. 8.18). As shown in the figure, latency critical queries still suffer from up to 5.1x QoS violation in Baymax-NP.

Even if a latency critical application is not PCI-e intensive, its occasional data transfer can be severely delayed by memcpy tasks from batch applications. For example, while less than 10% of GPU time is spent on PCI-e data transfer for *imc* and *face*, they still suffer from severe QoS violation due to the unmanaged and unpredicted PCI-e bandwidth contention in Baymax-NP.

Figure 8.18c shows that the accelerator utilization in Baymax and Baymax-NP are similar for most of the co-locations. This is mainly because existing emerging latency critical applications do not transfer data between CPU and GPU frequently, and the duration of their memcpy tasks is often less than 10 milliseconds (Fig. 8.11). In this case, the memcpy tasks in batch applications will not be delayed seriously and the accelerator utilization is not reduced seriously in Baymax compared with in Baymax-NP.

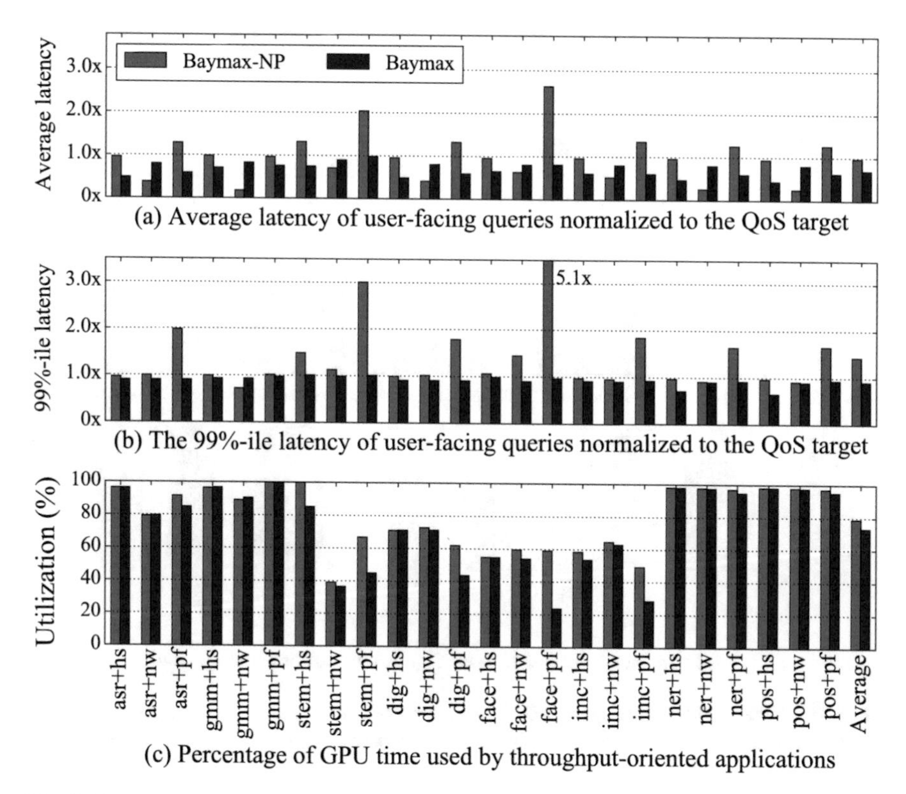

Fig. 8.18 Normalized average latency, 99%-ile latency of latency critical queries, and accelerator utilization when latency critical applications are co-located with PCI-e intensive batch applications. Different from Baymax, Baymax-NP does not mitigate PCI-e bandwidth contention through scheduling data transfer tasks

8.10.4 *Beyond Pair-Wise Co-locations*

To evaluate the robustness of Baymax in dealing with more complex co-location scenarios, we pick all the Rodinia benchmarks in Table 8.1 to form a mixture of batch applications, and co-locate them all with the latency critical applications from both Sirius suite and Tonic suite.

We report the normalized average latency and 99%-ile latency of latency critical queries, and accelerator utilization in this scenario in Fig. 8.19. As shown in the figure, Baymax is robust enough to increase the accelerator utilization while guaranteeing the QoS of latency critical applications. The average latency and 99%-ile latency of latency critical applications with Baymax and Baymax-NC are always within the QoS target as shown in Fig. 8.19a and b. On the contrary, Baymax-NP cannot satisfy the QoS of latency critical application (up to 1.6x QoS violation in terms of 99%-ile latency) due to the unawareness of PCI-e bandwidth contention. Compared with Baymax-NP, Baymax can achieve similar utilization improvement while satisfying

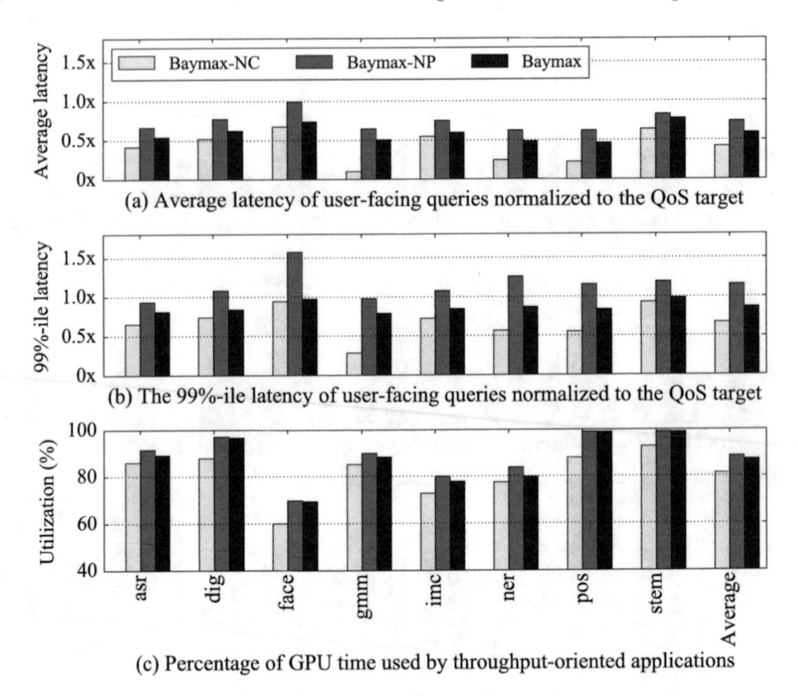

Fig. 8.19 Normalized average latency, 99%-ile latency of latency critical queries, and accelerator utilization when each latency critical application is co-located with all the batch applications

the QoS of all the latency critical applications. Compared with Baymax-NC, Baymax can further increase the average accelerator utilization from 81.2 to 87.4% as shown in Fig. 8.19c.

8.11 Summary

In this chapter, we discuss task scheduling techniques for non-preemptive accelerators. Limited by the existing GPU design, there is no open interface to schedule tasks that are already launched to the GPU. We therefore design a mechanism to schedule accelerator tasks on the CPU side. Based on the proposed scheduling mechanism, we further design and implement Baymax to improve the utilization of accelerator hardware while guaranteeing the required QoS target of latency critical time-sensitive applications.

Through real-system investigation, we find that four main factors affect the queuing delay and data transfer latency and thus the end-to-end latency of latency critical applications. These factors include the number of tasks in a latency critical query that indicates how many tasks could be delayed, the task execution order that decides which tasks may cause the delay for each latency critical task, the duration and occupancy of batch tasks that impact the queuing delay of each task in a latency critical query, as well as the PCI-e bandwidth contention that affects data transfer rate between host memory and accelerator memory.

Because key factors such as the task execution order and PCI-e bandwidth contention may change during runtime, an offline solution is not adequate. A runtime system, Baymax, that can dynamically monitor the accelerator and PCI-e bus, and schedule tasks accordingly is needed to maximize accelerator utilization while satisfying QoS of latency critical applications. **Baymax** composes of two parts: a *task duration predictor* and a *task scheduling engine*. The task duration predictor leverages novel models to predict the duration of tasks across different inputs. The task scheduling engine then intercepts and analyzes task launching function calls before passing control to the accelerator. Based on the precisely predicted task duration, Baymax schedules compute tasks issued to the accelerator. Meanwhile, Baymax limits the number of concurrent active data transfer tasks to mitigate PCI-e bandwidth contention. By scheduling tasks and managing PCI-e bandwidth, Baymax guarantees that QoS of latency critical applications is always satisfied regardless of the order of tasks issued by applications.

8.11.1 Chapter Highlights

The following highlights of this chapter could be of your interest:

1. **Design of a task scheduling mechanism to manage accelerator tasks**—We design a task scheduling mechanism that intercepts and schedule task invocations from co-located applications.
2. **Comprehensive analysis of QoS violation on non-preemptive accelerators**—We identify four key factors that significantly affect the end-to-end latency of latency critical applications when they are co-located with other applications. The analysis motivates the design of a task scheduling system based on precisely predicted task duration for accelerator co-locations.
3. **Design of online task duration prediction models**–We establish accurate and low-overhead models to estimate the duration of tasks on accelerators.
4. **Design of an online mechanism to mitigate PCI-e bandwidth contention**—We design a mechanism that monitors the realtime data transfer pressure on PCI-e bus and mitigates PCI-e bandwidth contention to eliminate QoS violation.

We implement Baymax runtime system combining all the above techniques. Baymax enables precise kernel duration prediction, QoS aware kernel scheduling, and PCI-e bandwidth contention aware data transfer management. Through evaluating Baymax with emerging latency critical workloads, we demonstrate the effectiveness of Baymax in eliminating QoS violation due to kernel interference and PCI-e bandwidth contention. Beyond the pair-wise co-locations, Baymax can improve the accelerator utilization by 87.4% without violating the QoS of 99%-ile latency for latency critical applications.

References

1. Alex Goldhammer and John Ayer Jr. Understanding Performance of PCI Express Systems. *Xilinx WP350, Sept*, 4, 2008.
2. Andrew Putnam, Adrian M Caulfield, Eric S Chung, Derek Chiou, Kypros Constantinides, John Demme, Hadi Esmaeilzadeh, Jeremy Fowers, Gopi Prashanth Gopal, Jordan Gray, et al. A Reconfigurable Fabric for Accelerating Large-scale Datacenter Services. In *the 41st International Symposium on Computer Architecture (ISCA)*, pages 13–24. ACM/IEEE, 2014.
3. Sunil Arya, David M Mount, Nathan S Netanyahu, Ruth Silverman, and Angela Y Wu. An Optimal Algorithm for Approximate Nearest Neighbor Searching Fixed Dimensions. Journal of the ACM, 45(6):891–923, 1998.
4. Chih-Chung Chang and Chih-Jen Lin. LIBSVM: A Library for Support Vector Machines. ACM Transactions on Intelligent Systems and Technology, 2(3):27, 2011.
5. Q. Chen, H. Yang, J. Mars, and L. Tang. Baymax: QoS Awareness and Increased Utilization for Non-Preemptive Accelerators in Warehouse Scale Computers. In *the Twenty-First International Conference on Architectural Support for Programming Languages and Operating Systems*, pages 681–696. ACM, 2016.
6. CUDA Nvidia. cuBLAS library. *Nvidia Corporation, Santa Clara, California*, 15, 2008.
7. Cui Yu, Rui Zhang, Yaochun Huang, and Hui Xiong. High-Dimensional KNN Joins with Incremental Updates. Geoinformatica, 14(1):55–82, 2010.
8. David Kirk et al. Nvidia CUDA Software and GPU Parallel Computing Architecture. In *the 6th International Symposium on Memory Management (ISMM)*, volume 7, pages 103–104. ACM, 2007.
9. Gareth James, Daniela Witten, Trevor Hastie, and Robert Tibshirani. *An Introduction to Statistical Learning*. Springer, 2013.
10. George AF Seber and Alan J Lee. *Linear Regression Analysis*, volume 936. John Wiley & Sons, 2012.
11. A. Ghodsi, M. Zaharia, B. Hindman, A. Konwinski, S. Shenker, and I. Stoica. Dominant resource fairness: Fair allocation of multiple resource types. In *NSDI*, 2011.
12. Glenn Elliott, Bryan C Ward, and James H Anderson. GPUSync: A Framework for Real-time GPU Management. In *the 34th Real-Time Systems Symposium*, pages 33–44. IEEE, 2013.
13. Glenn A Elliott and James H Anderson. Globally Scheduled Real-time Multiprocessor Systems with GPUs. Real-Time Systems, 48(1):34–74, 2012.
14. Haeseung Lee, Al Faruque, and Mohammad Abdullah. GPU-EvR: Run-time Event based Real-time Scheduling Framework on GPGPU Platform. In *Design, Automation and Test in Europe Conference and Exhibition (DATE)*, pages 1–6. IEEE, 2014.
15. Ivan Tanasic, Isaac Gelado, Javier Cabezas, Alex Ramirez, Nacho Navarro, and Mateo Valero. Enabling Preemptive Multiprogramming on GPUs. In *the 41st International Symposium on Computer Architecuture (ISCA)*, pages 193–204. ACM/IEEE, 2014.
16. Jason Jong Kyu Park, Yongjun Park, and Scott Mahlke. Chimera: Collaborative Preemption for Multitasking on a Shared GPU. In *International Conference on Architectural Support for Programming Languages and Operating Systems (ASPLOS)*, pages 593–606. ACM, 2015.
17. Jeffrey Dean and Luiz André Barroso. The Tail at Scale. Communications of the ACM, 56(2):74–80, 2013.
18. Johann Hauswald, Michael A. Laurenzano, Yunqi Zhang, Cheng Li, Austin Rovinski, Arjun Khurana, Ron Dreslinski, Trevor Mudge, Vinicius Petrucci, Lingjia Tang, and Jason Mars. Sirius: An Open End-to-End Voice and Vision Personal Assistant and Its Implications for Future Warehouse Scale Computers. In *the 20th International Conference on Architectural Support for Programming Languages and Operating Systems (ASPLOS)*. ACM, 2015.
19. Johann Hauswald, Yiping Kang, Michael A. Laurenzano, Quan Chen, Cheng Li, Ronald Dreslinski, Trevor Mudge, Jason Mars, and Lingjia Tang. DjiNN and Tonic: DNN as a Service and Its Implications for Future Warehouse Scale Computers. In *the 42nd Annual International Symposium on Computer Architecture (ISCA)*, pages 27–40. ACM/IEEE, 2015.

20. Kittisak Sajjapongse, Xiang Wang, and Michela Becchi. A Preemption-based Runtime to Efficiently Schedule Multi-process Applications on Heterogeneous Clusters with GPUs. In *the 22nd International Symposium on High-performance Parallel and Distributed Computing (HPDC)*, pages 179–190. ACM, 2013.
21. Luiz André Barroso, Jimmy Clidaras, and Urs Hölzle. The Datacenter as a Computer: An Introduction to the Design of Warehouse-Scale Machines. Synthesis Lectures on Computer Architecture, 8(3):1–154, 2013.
22. Nicola Jones. The Learning Machines, 2014.
23. Nvidia Multi-Process Service. https://docs.nvidia.com/deploy/pdf/CUDA_Multi_Process_Service_Overview.pdf.
24. Pedro Aguilera, Katherine Morrow, and Nam Sung Kim. QoS-aware Dynamic Resource Allocation for Spatial-multitasking GPUs. In *the 19th Asia and South Pacific Design Automation Conference (ASP-DAC)*, pages 726–731. IEEE, 2014.
25. Ping Xiang, Yi Yang, and Huiyang Zhou. Warp-level Divergence in GPUs: Characterization, Impact, and Mitigation. In *the 20th International Symposium on High Performance Computer Architecture (HPCA)*, pages 284–295. IEEE, 2014.
26. Profiler User's Guide. http://docs.nvidia.com/cuda/profiler-users-guide.
27. Sergio Barrachina, Maribel Castillo, Francisco D Igual, Rafael Mayo, and Enrique S Quintana-Orti. Evaluation and Tuning of the Level 3 cuBLAS for Graphics Processors. In *International Parallel and Distributed Processing Symposium (IPDPS)*, pages 1–8. IEEE, 2008.
28. Sharan Chetlur, Cliff Woolley, Philippe Vandermersch, Jonathan Cohen, John Tran, Bryan Catanzaro, and Evan Shelhamer. cuDNN: Efficient Primitives for Deep Learning. arXiv:1410.0759, 2014.
29. Shinpei Kato, Karthik Lakshmanan, Raj Rajkumar, and Yutaka Ishikawa. TimeGraph: GPU Scheduling for Real-time Multi-tasking Environments. In *USENIX Annual Technical Conference (ATC)*, pages 17–30. USENIX, 2011.
30. Shuai Che, Michael Boyer, Jiayuan Meng, David Tarjan, Jeremy W Sheaffer, Sang-Ha Lee, and Kevin Skadron. Rodinia: A Benchmark Suite for Heterogeneous Computing. In *IEEE International Symposium on Workload Characterization (IISWC)*, pages 44–54. IEEE, 2009.
31. Vincent Garcia, Eric Debreuve, and Michel Barlaud. Fast K Nearest Neighbor Search using GPU. In *Conference on Computer Vision and Pattern Recognition Workshops (CVPRW)*, pages 1–6. IEEE, 2008.
32. Vinicius Petrucci, Michael Laurenzano, John Doherty, Yunqi Zhang, Daniel Mosse, Jason Mars, and Lingjia Tang. Octopus-Man: QoS-driven Task Management for Heterogeneous Multicores in Warehouse-Scale Computers. In *the 21st International Symposium on High Performance Computer Architecture (HPCA)*, pages 246–258. IEEE, 2015.
33. Z. Wang, J. Yang, R. Melhem, B. Childers, Y. Zhang, and M. Guo. Quality of service support for fine-grained sharing on gpus. In *Proceedings of the 44th Annual International Symposium on Computer Architecture*, pages 269–281. ACM, 2017.
34. Z. Wang, J. Yang, R. Melhem, B. Childers, Y. Zhang, and M. Guo. Simultaneous multikernel gpu: Multi-tasking throughput processors via fine-grained sharing. In *High Performance Computer Architecture, 2016 IEEE International Symposium on*, pages 358–369. IEEE, 2016.

Part III
Summary and Discussion

Chapter 9
Summary and Discussion

Abstract In this chapter, we first introduce the guideline of designing the new task scheduling policies. Then, we give our perspectives on developing efficient and effective task scheduling techniques on various complex parallel architecture.

9.1 Guideline of Scheduling Technique Design

In this book, we have discussed emerging techniques used to improve the performance of task-based applications on various complex parallel architecture, such as MSMC architecture, NUMA-enabled architecture, AMC architecture, CPU+GPU heterogeneous architecture, heterogeneous Cloud, and Accelerators. As we can see from this book, different techniques are used for different parallel architectures. There is not an universal optimal technique that fits all the emerging parallel architectures. When a new parallel architecture is released, in order to achieve the best performance, we suggest to tune existing task scheduling techniques for the specific architecture accordingly.

In more detail, if a new parallel architecture is released, we suggest readers design new task scheduling techniques following three steps as follows.

1. Understand the hardware features of the architecture.
2. Based on the hardware features, we can identify the potential bottlenecks in the target architecture.
3. Modify existing task scheduling technique or develop a new task scheduling technique that attacks the bottlenecks which often result in the poor performance of parallel applications on the architecture.

In the following of this chapter, we introduce how we develop the task scheduling techniques for emerging parallel architectures following the above steps.

© Springer Nature Singapore Pte Ltd. 2017

Q. Chen and M. Guo, *Task Scheduling for Multi-core and Parallel Architectures*,
https://doi.org/10.1007/978-981-10-6238-4_9

9.2 Multi-socket Architecture

Following to the above three steps, let us first understand the hardware features of MSMC architecture. Compared with traditional single-socket multi-core architecture, the key feature of MSMC architecture is that the cores in the same socket share the last level cache, but cores in different sockets only shared the main memory. With this new feature, it is beneficial if a core reads the data into the shared cache while other cores in the same socket access the data as well.

Then, according to the above analysis, the potential bottleneck of MSMC architecture is the poor utilization of shared cache if tasks are scheduled randomly. With random scheduling, cores in the same socket often perform on different data, thus the data stored in the shared cache is not able to be reused. In this case, parallel applications, especially data-intensive applications, suffer from low performance due to the poor shared cache utilization.

Therefore, the CAB task scheduler in Chap. 3 attacks the poor shared cache utilization in multi-socket architecture. By improving the shared cache utilization through cache friendly task graph partition and bi-tier work-stealing policy, CAB significantly improve the performance of data-intensive applications on multi-socket architecture as shown in Sect. 3.8.

Although CAB is able to improve the performance of many data-intensive applications, it is still not perfect. For instance, it assumes neighbor tasks in the task graph share some data. Although many regular parallel applications follows the assumption but some other irregular parallel applications such as graph applications do not follow the assumption. It is still open to develop efficient and effective task scheduling technique for applications with irregular data access pattern on multi-socket architecture.

9.3 NUMA-Enabled Multi-socket Architecture

Compared with traditional multi-socket architecture, the key feature of NUMA-enabled multi-socket architecture is that the main memory is divided into memory nodes and each memory node is attached with a CPU socket. It is much faster for the cores in a socket to access data stored in its local memory node than remote memory nodes. With this new feature, it is beneficial if a core can already find the required from the local memory node instead of slower remote memory nodes.

According to the above analysis, besides the low shared cache utilization, another potential bottleneck of NUMA-enabled multi-core architecture is the large amount of remote memory accesses. With random task scheduling, a task is highly possible to be schedule to the socket where it has to access data from remote memory nodes. In this case, parallel applications, especially data-intensive applications, suffer from low performance due to the severe remote memory accesses.

Therefore, the LAWS task scheduler in Chap. 4 attacks both the poor shared cache utilization and low local memory accesses in NUMA-enabled multi-socket architecture. By improving the shared cache utilization and increasing local memory accesses through load-balanced data allocation, cache friendly task graph partition and triple-level work-stealing policy, LAWS significantly improve the performance of data-intensive applications on NUMA-enabled multi-socket architecture as shown in Sect. 4.10.

Although LAWS is able to improve the performance of many data-intensive applications on NUMA-enabled multi-socket architecture, it is not applicable for all the applications. For instance, it assumes each task in an application only processes a small portion of the whole dataset of the application. Although many regular parallel applications follows the assumption but some other irregular parallel applications such as graph applications do not follow the assumption. It is still open to develop efficient and effective task scheduling technique for applications with irregular data access pattern on NUMA-enabled multi-socket architecture.

9.4 Asymmetric Multi-core Architecture

Compared with traditional multi-socket architecture, the key feature of asymmetric multi-core architecture is that individual cores have different computational capabilities. With this new feature, it is beneficial to balance the workload across the asymmetric cores.

According to the feature of asymmetric multi-core architecture, the potential bottleneck of parallel applications on AMC architecture is the poor load balance. With traditional random task scheduling, it is highly possible that a long task is allocated to a slow core while a short task is allocated to a fast core. The example in Sect. 5.2 has shown that random task scheduling degrades the overall performance seriously.

Therefore, the AATS task scheduler in Chap. 5 attacks the poor load balancing problem in asymmetric multi-socket architecture. By perfectly balancing the workload through history-based task allocation and preference-based work-stealing, AATS significantly improve the performance of compute-intensive applications on asymmetric multi-socket architecture as shown in Sect. 5.10.

Although AATS is able to improve the performance of some compute-intensive applications, it is still not perfect. It assumes that tasks executing the same function in the current run have similar workloads. This assumption is not obeyed by all the applications. For instance, if all the tasks execute the same function but operate on data of different sizes, they have totally different workloads but are classified into the same task class in AATS. In this case, AATS operates the same as traditional random task scheduling and is not able to further balance the workload.

In addition, if there are dependencies between tasks, the history-based task allocation is not working. It is still open to develop new task scheduling techniques for more complex parallel applications on asymmetric multi-core architecture.

9.5 Heterogeneous CPU+GPU Architecture

Heterogeneous CPU+GPU architecture can also be viewed to be an asymmetric architecture where CPU is slow processing element and GPU is fast processing element. Different from the targeted single-ISA AMC architecture in Chap. 5, CPU and GPU have different ISAs. One key feature of CPU+GPU architecture is that different applications have different speedup ratios on the GPU compared with CPU, because the applications have various characteristics. In addition, a GPU performs poor when a GPU kernel is too small to fully utilize all its SMs.

According to the above analysis, it is not trivial to find an universal optimal workload allocation for all the parallel applications offline. The potential bottleneck of CPU+GPU architecture is the poorly balanced workload and the poor performance of GPU due to the small workload of GPU kernels. If the workload is not balanced, the overall performance of a parallel application is determined by the slowest side of the CPU+GPU architecture. If fine-grained task scheduling is used to balance the workload, GPU performs poor in this case and further results in the poor performance as well.

Therefore, the HATS task scheduler in Chap. 6 attacks the above two bottlenecks in the heterogeneous CPU+GPU architecture. By balancing the whole workload across CPU and GPU while maximize the task granularity through asymptotic profiling, HATS significantly improve the performance of applications on CPU+GPU architecture as shown in Sect. 6.5.

HATS task scheduler is also not perfect for heterogeneous CPU+GPU architecture, because it potentially assumes that the workload of a task increases linearly with the size of its data set. This assumption is not always correct in complex parallel applications such as algorithms of sparse matrix. It is still open to develop effective task scheduling techniques that can balance workload across CPU and GPU for real-system complex parallel applications.

9.6 Heterogeneous Cloud Platform

The key feature of heterogeneous Cloud platform is that the nodes have different processing speeds for map tasks and reduce tasks. Obviously, the performance of MapReduce is often seriously damaged by a few straggler tasks that complete far behind the other tasks in heterogeneous Cloud platform. The emerging solution of speeding up the straggler tasks is to launch backup tasks for them on fast nodes. However, current least progress policy used to identify straggler tasks is not able to correctly identify the actual straggler tasks thus fails to maximize the performance of MapReduce applications.

According to the above analysis, one of the potential bottleneck of MapReduce applications on heterogeneous Cloud platform is the in-accuracy of identifying actual straggler tasks. In a wrong task is identified to be a straggler task, the actual straggler tasks will not be speeded up and the backup task of the wrong straggler task wastes resources. In this case, the actual straggler tasks and the wasted resources together result in the poor performance of MapReduce applications on heterogeneous Cloud platform.

Therefore, the SAMR scheduler in Chap. 7 attacks the problem of poor accuracy of identifying real straggler tasks. By accurately calculating the remaining time of all the tasks leveraging the weights of different phases of a map/reduce task, SAMR significantly improve the performance of Map-Reduce applications on heterogeneous Cloud architecture as shown in Sect. 7.5.

Although SAMR is able to improve the performance of many Map-Reduce applications, it is still not perfect because straggler tasks are not the only bottleneck of a MapReduce application on heterogeneous Cloud platform. For instance, it is beneficial to balance the workload across heterogeneous nodes and reduce network congestion in the Cloud platform. It is still open to develop efficient and effective task scheduling technique for emerging MapReduce applications on heterogeneous Cloud platform.

9.7 Non-preemptive Accelerator Architecture

Many emerging accelerators, such as GPGPU and FPGA, are non-preemptive. For these non-preemptive accelerators, the key feature is that a newly submitted kernel is not able to preempt the accelerator but can only start to run after the current running kernel completes. Although the latest GPU Nvidia P100 already supports preemption, the overhead of preemption is too large to be used in real system. With this feature, when a latency-critical application is co-located with other applications on the same non-preemptive accelerator, the latency-critical application can be severely delayed by the co-located applications.

According to the above analysis, the bottleneck of guaranteeing the QoS of latency-critical applications on accelerator at co-location is the non-preemptive feature. For instance, if a long kernel of an application is running on the accelerator, all the following kernels are delayed until the long kernel completes. In this case, the QoS of latency-critical applications cannot be satisfied.

Therefore, the Baymax scheduler in Chap. 8 attacks the QoS violation problem of latency-critical applications at co-location on non-preemptive accelerator. By precisely predicting the duration of every kernel, reordering all the kernels accordingly, mitigating PCI-e bandwidth contention, Baymax is able to greatly improve the accelerator utilization while guaranteeing the QoS of latency-critical applications as shown in Sect. 8.10.

At last, I want to emphasize again that there is no universal scheduling policy that can work perfectly for all parallel architectures. Therefore, as parallel architectures become increase in complexity, dynamic task scheduling techniques will need to be optimized for their specific features in the three steps described in this chapter.

Glossary

Shared Memory Parallel Architecture A computer architecture that consists of multiple processing elements and the main memory is shared by all the processing elements.

Distributed Memory Parallel Architecture A computer architecture that consists of multiple processing elements and the main memory is distributed to different nodes. Each processing element is only able to access part of the main memory attached with its node.

Multi-socket Multi-core Architecture (MSMC) It is a kind of shared memory parallel architecture, in which multiple CPU chips are integrated into a single node and each CPU chip has multiple cores with a shared last-level cache. Each CPU chip is plugged into a socket.

Asymmetric Multi-core Architecture (AMC) It is a kind of shared memory parallel architecture that consists of a mix of fast cores and slow cores.

GPGPU General Purpose GPU that can be used to process general applications besides graph applications.

IPC Instruction-Per-Cycle. It is used to measure the processing speed of an application on a computer.

Makespan The actual time of processing an parallel application (From the time that the application is launched to the time that the application terminates).

SM Streaming Multiprocessor in GPGPU.

Manual Task Scheduling Tasks are scheduled by programmers manually in order to balance the workload between threads/processes for the good performance.

Automatic Task Scheduling Tasks are scheduled automatically at runtime in order to balance the workload between threads/processes for the good performance.

TRICI Task relocation incurred cache interference problem.

Cache-Aware Scheduling Schedule tasks when considering the impact from cache usage.

Locality-Aware Scheduling Schedule tasks when considering where the data of a task is stored.

Task Graph The execution of a task-based parallel application can be expressed to be the traversal of a task graph, which is a Directed Acyclic Graph (DAG).

© Springer Nature Singapore Pte Ltd. 2017
Q. Chen and M. Guo, *Task Scheduling for Multi-core and Parallel Architectures*,
https://doi.org/10.1007/978-981-10-6238-4

In the task graph, the nodes represent the tasks and the edges correspond to the dependence relationship among the tasks.

SOID The size of data involved by a task during its execution.

NUMA Non-Uniform Memory Access. In MSMC architecture, the main memory is divided into multiple memory nodes and each node is attached to the socket of a chip. The memory node attached to a socket is called its local memory node and those that are attached to other sockets are called remote memory nodes. The cores of a socket access its local memory node much faster than the remote memory nodes.

First touch strategy A strategy of data allocation in NUMA memory. If a chunk of data is first accessed by a task that is running on a core of the socket ρ, a physical page from the local memory node of ρ is automatically allocated to stored the data.

CF subtree It is a subtree of the whole task graph of an application. The data involved by all the tasks in the subtree is small enough to fit into the last level cache.

Cache Miss If a core cannot find the needed data in the cache, a cache miss happens. In this case, the core tries to access data from lower level of cache or memory.

Local Memory Access A core read the needed data from the memory node attached to its socket.

Parent-first policy It is a task generation policy in work-stealing, where a core continually executes the parent task after spawning a new task.

Child-first policy It is a task generation policy in work-stealing, where a core continually executes the spawned new task once the child is spawned.

Cilk2c It is the source-to-source compiler of MIT Cilk.

Straggler task/socket It is the task/socket that needs the longest time to complete its work. It can significantly degrade the performance of a parallel application.

C-group A group of cores that operate at the same speed and have the same performance in asymmetric multi-core architecture.

Task class A task class is comprised of a group of tasks that run the same function.

LATE policy The longest approximate time to end policy. The task that has the longest approximate time to its end is considered to be the straggler task.

Speculative Execution With speculative execution, if a map/reduce task is considered to be the straggler task, the system launches a backup of the map/reduce task speculatively on a fast node. Either the original task or the backup task completes, the task is considered to be complete successfully.

Preemptive A hardware (e.g., CPU, GPU, FPGA) is preemptive when a newly submitted task t can stop the currently running task and preempts the hardware. Similarly, a hardware is non-preemptive if a task can use the hardware only when the current task running on the hardware completes.

QoS Quality-of-Service (QoS) is an advanced feature that prioritizes one application to minimize the cases that the application completes after the deadline.

MPS MPS (Multi-Process Service) scheduling is a technique proposed by Nvidia that enables concurrent sharing of a GPU among multiple applications.

Memcpy Copy data between the main memory and the global memory of the accelerator through PCIe bus.

ANN It is a machine learning algorithm called $ApproximateNearestNeighbor$.

KNN It is a machine learning algorithm called $K - NearestNeighbor$. K is an integer configured by the user.

Printed in the United States
By Bookmasters